Professional Practice in Governance and Public Organizations

The *Professional Practice in Governance and Public Organizations* series features cutting-edge insights and practical guidance for professionals in the areas of economics, politics, public policy, public administration, and international organizations. With concise, accessible volumes, the series explores the latest developments in governance, public law-making, organizational and political strategies, institutional policies, policy instruments, public management, and finance. Core topics include decision-making, leadership, and the transformative impact of digitalization.

Each book is authored by practitioners, experts, and leading authorities from public and international organizations, think tanks, and NGOs, ensuring a blend of theoretical depth and real-world applicability. While designed primarily for professionals in these fields, the series also serves as an invaluable resource for students of economics, political science, public policy, and public administration, equipping them with practical guides for their future careers.

Remi Chandran

The Wildlife Enforcement Monitoring System

From Silos to Synergy in Environmental Governance

Springer

Remi Chandran
Remote Sensing Technology Center of Japan
Tokyo, Japan

ISSN 2731-9776 ISSN 2731-9784 (electronic)
Professional Practice in Governance and Public Organizations
ISBN 978-3-031-97083-2 ISBN 978-3-031-97084-9 (eBook)
https://doi.org/10.1007/978-3-031-97084-9

This Springer imprint is published by the registered company Springer Nature Switzerland AG
The registered company address is: Gewerbestrasse 11, 6330 Cham, Switzerland

Preface

Through all the ups and downs of my life, I have stayed committed to one goal: creating a meaningful way for four key groups—governments, UN experts, NGOs and scientists—to share information and work together on wildlife conservation and enforcement. That goal led to the creation of the Wildlife Enforcement Monitoring System (WEMS)—a tool, though not alive in itself, designed to support these groups in the shared mission of tackling wildlife crime.

But building WEMS was not an easy journey. Even though these groups all showed interest in working across borders, agreeing on a shared system proved difficult. Each had its own ideas, rules and ways of working, which made it hard to find common ground. In this book, *Silos to Synergy: The Wildlife Enforcement Monitoring System and Paradigm Shifts in Transnational Data Sharing,* I describe this challenge as a missing 'boundary arrangement'—a shared space where different conflicting perspectives can meet without losing their identity. Without this kind of shared understanding, it is very hard to solve the long-standing issues in wildlife enforcement information-sharing.

What gives me the confidence to write about this is not just my understanding of the system, but my experience working closely with the people and organisations involved. I have been part of many of these groups myself, and that experience taught me something important: good work should always aim to bring about positive change. But I also learned that it can be difficult—sometimes even impossible—to question a global process deeply while still being part of it. This is perhaps why many UN reports are written in vague or soft language, making it easier for other organisations to offer their own interpretations.

When I was a student, I was taught to look for the 'right' answers—clear and simple solutions. But once I started working on real-world problems, I saw that things are rarely so straightforward. Complex issues don't have fixed answers; they depend on history, context and the people involved. When WEMS faced resistance, I kept going. I adjusted the project each time it was challenged—what I now call *policy-oriented learning.* With every pushback, I looked at what wasn't working and tried to rebuild the project in a better way. The approach described in *Silos to Synergy* is just one possible way forward—it's not the final answer.

Over time, I also came to see that the lessons from this journey apply far beyond wildlife enforcement. The ideas explored here—about information-sharing, institutional conflict, power dynamics and policy learning—are relevant to many fields. Whether in environmental governance, public health, law enforcement, international relations or negotiations, the need to bridge differences and build common ground remains the same.

Just as importantly, I have come to realise how critical timely information-sharing is—not only for protecting fragile ecosystems but also for saving human lives. In an increasingly connected world, delays in communication and action can have consequences that cross borders and affect both nature and society. The need to act together, and to act swiftly, has never been more urgent.

In recent times, I have begun to feel the quiet toll that this long and often solitary journey has taken—on both body and mind. The effort has been tireless, and at times exhausting. As I reflect now, I sense that my chapter in this story may be drawing to a close.

But the journey itself must go on—led by new frames of thinking, inspired by fresh ideas and carried forward by a new generation of actors who are ready to take up this responsibility with courage and clarity.

It is with that understanding—and with quiet hope—that I place this work in your hands.

Remi Chandran

Acknowledgements

Silos to Synergy is the culmination of a journey that spans two decades and several continents, marked by countless challenges and moments of profound learning. It reflects my work as the Project Lead for the Wildlife Enforcement Monitoring System (WEMS) and the people whose unwavering support, encouragement and belief in this project made it possible.

First, I owe my deepest gratitude to the WEMS initiative team—scientists, practitioners and experts from all over the world. I am particularly thankful to the late Prof. Hans van Ginkel, former Rector of the United Nations University (UNU), whose bold vision and commitment to human geography were key reasons he supported WEMS in 2003. Without his leadership, this project would never have materialised. I also want to thank Prof. Padmanabhan Krishnan (Director of Research, Oracle), Mr. Khoi Nguyen (WEMS CTO and NTT Data), Mr. Masayuki Sakamoto (Executive Director, JTEF), Prof. Govindan Parayil (former Vice Rector, UNU), Mr. Max Bond (former Vice Rector, UNU) and Mr. Kaname Ikeda (President, RESTEC)—each one a pillar of this initiative—whose support and belief in this mission have been vital every step of the way.

The evolution of WEMS in 2007 was shaped by the invaluable guidance of Ms. Elizabeth Maruma Mrema (Assistant Secretary-General of the United Nations and Deputy Executive Director, UNEP) and Mr. John Sellar (former CITES enforcement officer), whose expertise in legal and policy issues helped transform our early ideas on government-led systems into something truly impactful. Special thanks to Jack Dangermond, President of ESRI, for his steadfast support of the initiative, as well as to Mike Gould, Peter Eredics and the entire ESRI team whose partnership has been crucial.

Silos to Synergy is an adaptation of my PhD thesis, 'If Wildlife Enforcement Monitoring System is a solution, what is the problem', for which I am endlessly grateful to my PhD supervisors—Prof. Yola Georgiadou, Prof. Robert Hoppe and Prof. Walter De Vries—whose mentorship was the foundation on which this book stands. Prof. Robert Hoppe's deep and invaluable policy knowledge was critical in shaping the policy processes discussed in this book, as well as the development of WEMS itself. His rigorous approach to policy analysis, combined with his ability to understand and navigate complex governance structures, enriched the work in ways that I will always carry with me. Prof. Yola's belief in me, despite the unconventional nature of my research, has been a beacon throughout. Her efforts in facilitating collaborations that bridged the gap between academic inquiry and fieldwork are woven into every page of this book. Prof. Walter's guidance on African governance and digital systems was invaluable—his calm wisdom kept me grounded through the complexities of this research.

The fieldwork for this book was made possible by so many extraordinary individuals across India, Thailand, Japan, Kenya, Tanzania and Uganda. My heartfelt thanks to Ms. Klairoong Poonpon (Specialist at the Department of National Parks, Wildlife, and Plant Conservation (DNP)), Mr. Samir Sinha (TRAFFIC India), Mr. Vinod Kumar (PCCF, Government of Haryana), Mr. S.P. Negi (Deputy Director General of Forests, MOEFCCC), Mr. Vivek Menon (WTI), the late Mr. Ashok Kumar (WTI), Ms. Kumi Togawa (JTEF) and all the experts at WWF, TRAFFIC and WCS. I also owe much to Mr. Edward Phiri, Mr. Bonaventure Ebayi and the dedicated team at the Lusaka Agreement Task Force for their tireless support in Sub-Saharan Africa. In Tanzania and Uganda, I remain deeply grateful to the officials at TANAPA and the Uganda Wildlife Authority for their openness and commitment to conservation.

A special mention goes to Dr. Ng Chong (Director, UNU C3), Mr. Khoi Nguyen (WEMS CTO) and Shankar Prakash (WTI)—who provided technical expertise and coordination that brought vital perspectives to the development of WEMS and I am forever thankful for their collaboration. My sincere appreciation also goes to the teams at CITES, UNEP, INTERPOL and the World Customs Organization, whose assistance was invaluable.

To my family, words cannot express the depth of my gratitude. To Yuki, your quiet strength has been the steady anchor that allowed me to pursue this work. Every chapter of this book carries the mark of your unwavering support. To Akash, your patience, understanding and unconditional belief in me have been my greatest source of inspiration—this book is as much yours as it is mine.

This book is dedicated to my parents. To my father, whose vision and values set me on this path, and to my mother, whose love and sacrifices have been the quiet force behind every achievement in my life. You are both with me in every word.

Lastly, I extend my heartfelt thanks to my dear childhood friend Shanuj Vayot, whose editorial insight and friendship were a guiding light through this process, and to Samuel Thomas for his invaluable intellectual support. My heartfelt gratitude also extends to the European Union Erasmus Mundus Programme, the United Nations University, and the Governments of India, Kenya, Tanzania, and Thailand, whose generous support and logistical assistance made this journey possible.

Finally, I thank Lorraine Klimowich, Senior Editor for Economics, Political Science, and Public Administration at Springer Nature, and her team for accepting to publish the book.

Prologue

Langdon Winner, in his essay 'Do Artifacts Have Politics' (Winner, 1980), suggests: 'No idea is more provocative in controversies about technology and society than the notion that technical things have political qualities.' He argues that machines, structures and systems of modern material culture can be evaluated not only for their efficiency, productivity and side effects, but also for the ways in which they embody specific forms of power and authority. Winner further claims that the adoption of a given technical system necessitates the creation and maintenance of particular social conditions as the operating environment of that system. For instance, an automobile, even though it may meet all functional requirements for operation, still requires legal approval (which defines its body structure and speed) before it can be used as a means of transport.

My experience with technology adoption, specifically in the development and implementation of a geospatial decision support system—the Wildlife Enforcement Monitoring System (WEMS)—made Winner's essay particularly enlightening. WEMS was developed by the United Nations University (UNU)[1] within the broader vision of 'open access', where information is regarded as a public good that, rather than depleting the common stock, enriches it to address complex global problems. Prominent figures such as Tim Berners-Lee,[2] Robert King Merton[3] and Elinor

[1] The United Nations University (UNU) was established in 1973. It functions as a think tank of the United Nations and carries out research on the pressing global problems faced by UN member states (UNU Charter). For this purpose, UNU provides academic freedom and autonomy to its scholars in advancing its objectives (Article II of UNU Charter).

[2] Tim Berners-Lee calls for open source information and software in his book *Weaving the Web: The Original Design and Ultimate Destiny of the World Wide Web by its Inventor*. Harper Business, 2000, ISBN 978-0-06-251587-2, pp. I–IX, 1–246.

[3] Robert King Merton advocated as early as 1942 that the results of research should be freely accessible to all. 'Each researcher must contribute to the "common pot" and give up intellectual property rights to allow knowledge to move forward.' See more at: http://access.okfn.org/2015/04/30/open-access-to-research-data-timeline Access date 21/10/2015

Ostrom[4] have all advocated for open data and public access, proposing that citizens should have the ability to judge issues concerning their well-being based on openly available information.

Inspired by the work of these 'open access' advocates, and in an attempt to address complex environmental problems, I, as part of the UNU team, envisioned a world where WEMS could support compliance with and enforcement of Multilateral Environmental Agreements (MEAs). Although the infrastructural design of WEMS was tailored to measure non-compliance with the Convention on International Trade in Endangered Species of Wild Fauna and Flora (CITES) by compiling and sharing information on poaching, illegal logging and the illegal trade of flora and fauna (collectively termed 'wildlife crime'), the UNU team anticipated that WEMS could facilitate broader transboundary cooperation. The tool was conceptualised as a single-window, open-information-sharing platform connecting multiple stakeholders, including governments, scientists, NGOs, local communities and industry. Furthermore, such a common platform was expected to inspire other MEAs, including the Convention on Biological Diversity (CBD), the United Nations Convention on the Law of the Sea (UNCLOS), the Convention on Migratory Species (CMS) and the Basel Convention (which regulates transboundary movements of hazardous waste and their disposal).

Understanding that certain conservation NGOs were well qualified and had tacit knowledge on wildlife enforcement matters, UNU decided to develop WEMS as a tool that could compile NGO-based information, assuming that the information generated from the platform would assist all relevant stakeholders in decision-making processes. Surprisingly, when the prototype was planned to be showcased in March 2006, the CITES Secretary-General objected to the information collection process inherent in the WEMS prototype. The apparent mismatch in the objection was that, while on the one hand, CITES was preaching that governments and civil society should provide as much *information* as possible on illegal trade so that CITES could take effective steps to prevent it; on the other hand, the administration of the Convention, through the CITES Secretary-General, stated that there were certain rules for the information collectors. In an apparent twist to the previous objective of WEMS, in 2007, the WEMS team restructured the project, excluding NGO participation, and decided to work directly with governments. There was also an attempt to implement WEMS in India, but again due to institutional politics among two different agencies within the Ministry of Environment of India, WEMS could not be implemented. WEMS was finally adopted by Kenya, Uganda, Tanzania and Congo through the assistance of the Lusaka Agreement Task Force (LATF). The information within WEMS became restricted to enforcement agencies and selective

[4]Elinor Ostrom was in support of free and open source information. According to her, bureaucrats sometimes do not have the correct information, while citizens and users of resources do. Extracted from 'Elinor Ostrom and The Digital Commons', *Forbes*; see http://www.forbes.com/2009/10/13/open-source-net-neutrality-elinor-ostrom-nobel-opinions-contributors-david-bollier.html Access date 21/10/2015

scientific partners within the WEMS consortium. Despite efforts to redesign WEMS to gain a larger acceptance, it never got fully endorsed by the CITES Secretariat.

In short, although WEMS was developed to find a solution to a problem, it was enmeshed in a web of obstacles. Chapter 1 of this book 'If WEMS is the solution … what is the problem?' is drawn from this apparent mismatch.

Contents

About the Author

Remi Chandran (PhD) is an environmental governance expert and professional policy negotiator with over 30 years of experience at the intersection of public policy, geospatial technology, and sustainability. In 2003, while at United Nations University he pioneered the Wildlife Enforcement Monitoring System (WEMS)—a transboundary geospatial tool in collaboration with Asian Conservation Alliance and ESRI Corporation and, later implementing it as regional tool for Lusaka Agreement member states in Africa (WEMS) Initiative revolutionized enforcement information systems by bringing the academic community into a space traditionally dominated by enforcement agencies—setting a global precedent for data transparency and collaborative governance.

Known for his honesty, integrity, and frank approach, Remi has earned the trust of several governments across Africa and Asia. He currently serves as a Senior Expert at the Remote Sensing Technology Center of Japan (RESTEC), where he works on geospatial strategies for climate action, biodiversity conservation, forest monitoring, and digital public infrastructure.

He also serves as an Advisor to the Lusaka Agreement Task Force in Nairobi, the UN Maps Programme at the United Nations Global Service Center (UNGSC) in Brindisi, Italy, and as a Visiting Researcher at the National Institute for Environmental Studies (NIES) in Japan.

Remi holds a PhD in Public Policy and Environmental Governance from the University of Twente, Netherlands, and an MSc in Agriculture from University College Dublin, Ireland. His career reflects a deep commitment to bridging data, diplomacy, and design thinking—developing systems-based approaches to today's pressing environmental challenges.

This book, Wildlife Enforcement Monitoring System, captures a significant part of Remi's journey in bringing innovation to the world of enforcement information sharing—challenging conventional boundaries and reshaping how knowledge and power intersect in environmental governance. Remi is a recipient of the Irish government bilateral fellowship (1995-97), the European Union Erasmus Mundus fellowship (2010) and the UNU-IAS PhD Fellowship (2012-2014).

Acronyms and Abbreviations

ACA	Asian Conservation Alliance
ACF	Advocacy Coalition Framework
CAMPFIRE	The Communal Areas Management Program for Indigenous Resources
CBD	Convention on Biological Diversity
CITES	Convention on International Trade in Endangered Species of Wild Fauna and Flora
COP	Conference of the Parties
CMS	Convention on the Conservation of Migratory Species of Wild Animals
ESRI	Environmental Systems Research Institute
EU-TWIX	European Union Trade in Wildlife Information eXchange
FAO	Food and Agriculture Organization of the United Nations
GATT	General Agreement on Tariffs and Trade
ICPO-Interpol	International Criminal Police Organization
ICCWC	The International Consortium on Combating Wildlife Crime
IIASA	The International Institute for Applied Systems Analysis
IPBES	Intergovernmental Science-Policy Platform on Biodiversity and Ecosystem Services
IS	Information Science
IUCN	IUCN—The World Conservation Union
IUCN/SSC	IUCN Species Survival Commission
IFAW	International Fund for Animal Welfare
LATF	The Lusaka Agreement Task Force
MEA	Multilateral Environmental Agreement
MONUSCO	United Nations Organization Stabilization Mission in the Democratic Republic of the Congo
MIKE	Monitoring the Illegal Killing of Elephants
NCB	National Central Bureaus
nCEN	National Customs Enforcement Network
NGOs	Non-governmental organisations

OSESG	UN Office of the Special Envoy for the Great Lakes Region
PAPPS	Public Administration and Political Policy Science
RILOS	Regional Intelligence Liaison Office
SDGs	Sustainable Development Goals
SC	Standing Committee
SSN	Species Survival Network
TRAFFIC	Trade Records Analysis of Flora and Fauna in Commerce
TWI2050	The World in 2050 project
UNEP	United Nations Environment Programme
UNEP-WCMC	UNEP World Conservation Monitoring Centre
UNFCCC	United Nations Framework Convention on Climate Change
UNGSC	United Nations Global Service Centre
UNODC	United Nations Office of Drugs and Crime
UNU	United Nations University
WEMS	Wildlife Enforcement Monitoring System
WCO	World Customs Organization
WTO	World Trade Organization
WWF	World Wide Fund for Nature

List of Figures

List of Tables

Chapter 1
If WEMS Is the Solution … What Is the Problem?

1.1 A Journey Through Science, Technology and Politics

Wildlife Enforcement Monitoring System (WEMS) was designed as a solution to address data gaps in addressing illegal wildlife trade. However, its development quickly became enmeshed in a web of unforeseen obstacles. This chapter unpacks that paradox—showing how a solution intended to meet a clear need ended up exposing deeper, more complex issues at the crossroads of technology, communication and system design. In the context of open data systems, one can further infer that information collectors inevitably rise to prominence when users become selective about their sources. This selectivity fosters an information bias, where users' trust is shaped by the perceived credibility of the source itself. Similar observations were made by Star and Ruhleder (1996) while working on a large-scale custom software effort, the Worm Community System (WCS), a collaborative system designed for a geographically dispersed community of geneticists. Star and Ruhleder (1996) figured out that no one was really using the system in spite of the fact that they were all claiming to be. Star and Ruhleder (1996) used Bateson's theory as a sensitising device and analysed the ways in which communicative processes were entangled in the development of ICT infrastructure by citing three levels of issues. They classified these issues in the context of 'signing on' and 'hooking up' to WCS, and discussed each level, with respect to the worm community and WCS, as follows:

- *First-level issues:* Primarily involving issues surrounding the 'finding out' and 'figuring out' of the installation and use of a system.
- *Second-level issues:* Issues formed as a result of unforeseen contextual effects, such as resistance to a programme or application, or lack of information on how to hook up to a system.

R. Chandran, *The Wildlife Enforcement Monitoring System*, Professional Practice in Governance and Public Organizations,
https://doi.org/10.1007/978-3-031-97084-9_1

- *Third-level issues:* These issues are more related to multiple meanings of information and network externalities which emerge in relation to competition, power and the role of secrecy.

When contextualising the work of Star and Ruhleder (1996) to the information usage pattern within CITES, observations can be made on the scientists working on CITES information-sharing-related research by distinguishing first-, second- and third-level issues, and classifying them as technical (first-level), governance (second-level) and political (third-level) issues. For instance, 'Boosting CITES', an article in the policy forum of *Science* (Phelps et al., 2010), mentions that CITES has already enhanced data sharing and analysis through collaborations with NGOs and partnerships, such as the Wildlife Enforcement Monitoring System.[1] The article lauded CITES for improving Party compliance and science-based decision-making despite political sensitivities (third-level issues) and chronically restricted funding, but singled out technical challenges (first-level issues) related to systematic and standardised data collection, rigorous data analysis and flexible research methods (Phelps et al., 2010). Moreover, the article and the responses it provoked in the policy forum of *Science* identified second- and third-level issues such as: governance and political challenges within CITES; lack of internal and external checks and balances; the high incentives for biased analyses and misreporting; and the lack of independence of CITES National Management Authorities from their advisory Scientific Authorities as well as the lack of public access to official parties' submissions to CITES (ibid.).

In this research, second- and third-level issues are defined as the policy-related and political challenges that impact the implementation of Information and Communication Technologies (ICT) within governmental and organisational contexts. These levels of challenges often extend beyond technical or operational concerns, encompassing broader institutional, regulatory and governance dynamics. Second-level issues typically involve the formulation and alignment of policies, legal frameworks and standards that guide ICT adoption. In contrast, third-level issues are rooted in political interests, power relations, stakeholder negotiations and broader socio-political environments that influence decision-making processes. The following section critically examines how these interconnected policy and political dynamics shaped the development of the Wildlife Enforcement Monitoring System (WEMS), highlighting the complex interplay between governance structures, institutional mandates and stakeholder agendas.

[1] It should be noted that WEMS was not endorsed by CITES; it had objected to the NGO-based enforcement information collection process within WEMS-Asia. However, CITES had approved WEMS as a potential tool which can support enforcement information-sharing. The later part of the book explains the reasons for the objection.

1.2 Contextualising the Development of WEMS

As discussed in the preceding section, a central aim of this research was to reveal the inherent paradox within WEMS—a paradox primarily rooted in secondary and tertiary concerns—alongside the modifications implemented to introduce the system. This entailed a thorough investigation into the reasons behind the failure of information-sharing enforcement within CITES and the challenges associated with the development of WEMS. The objections raised by the CITES Secretariat regarding WEMS, coupled with its eventual adoption in Africa, also warranted detailed analysis. To achieve this, it was necessary to situate these issues within the frameworks of Information Science (IS) and Public Administration and Political Policy Science (PAPPS), as illustrated in Fig. 1.1.

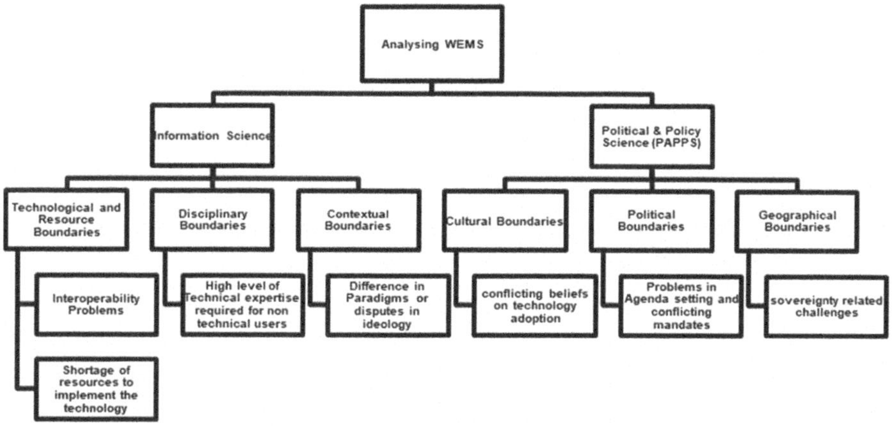

Fig. 1.1 Contextualises the development of WEMS within IS and PAPPS
The above figure contextualised the development of WEMS within the broad domains of IS and PAPPS. It could be observed that a distinct boundary existed between the IS perspective and the PAPPS perspective. The information science perspective typically addressed first- and second-level issues, but not third-level ones; whereas the policy/politics perspective engaged with third- and second-level issues. However, it should be noted that both IS and PAPPS dealt with governance (second-level issues). The relationship between IS and PAPPS, as outlined in Fig. 1.1, is briefly explained in the following sections

1.2.1 The Information Science (IS) Perspective

Within the field of information science, several authors had studied 'information systems' in relation to their organisational and institutional environments, particularly when implemented in developing countries. According to Avgerou (2008), ICT projects in developing countries were often deprived of resources or lost political commitment; they were poorly maintained and, consequently, technologically as well as functionally degraded. Orlikowski and Barley (2001) attributed failure to the lack of integration between IS research and knowledge from other disciplines. As Orlikowski and Barley argued, understanding and guiding techno-social developments required knowledge of technological systems, social processes and their interactions.

IS scholars often attributed the failure of ICT projects to a lack of resources or knowledge, occasionally exploring the role of power and politics in the design and implementation of information systems (see Richter, 2014). However, most IS scholars failed to delve deeper into the political and power structures that influenced ICT implementation. According to Orlikowski and Barley (2001), disciplinary boundaries continued to pose an obstacle to interdisciplinary research. As they succinctly stated: 'Unfortunately, the boundaries that we have drawn around our disciplines currently hamper the development of a more integrated approach' (Orlikowski & Barley, 2001, p. 158).

On the other hand, the influence of politics and organisational policies on information systems was often well understood in the public administration and political policy science (PAPPS) literature (see Snellen et al., 2012). The influence of politics and organisational policies was further elaborated in the following section.

1.2.2 The Public Administration and Political Policy Science (PAPPS) Perspective

In instances where information systems were managed or used by governments, failures sometimes went beyond technical issues. Contextual factors, as noted by Star and Ruhleder (1996), often played a significant role: "*Information infrastructure is not a substrate which carries information on it, or in it, in a kind of mind–body dichotomy. The discontinuities are not between system and person, or technology and organisation, but rather between contexts*" (Star & Ruhleder, 1996, p. 118).

It should be noted that information use by governments occurred at different levels (third-level issues) and in various policy contexts (second-level issues) within or during a decision-making process. When information was presented for discussion, its use often underwent various twists and turns, becoming embedded within a specific context, with the convenience of its usage ultimately defining its application. An unrefuted example of this can be seen in the European Spatial Development Perspective (ESDP) (detailed in Sect. 1.4), during the development of policy maps. This argument was further reinforced by the work of Weible et al. (2010), who outlined different strata of the use and impact of 'information' in the policy cycle, including:

(i) Identifying problems in agenda setting.
(ii) Forecasting impacts in policy formulation.
(iii) Identifying alternatives in policy adoption.
(iv) Monitoring impacts in implementation and evaluation.

Earlier studies by Dunn et al. (1981) stressed that unless information about activities and consequences was usable by the persons or agencies whose performance was being assessed, it would be perceived as irrelevant. However, the true reasons behind the objection or rejection of information by policymakers were often difficult to determine, making it crucial to understand the policy problem. Hoppe (2011) emphasised the need to identify the 'real' problem—the 'problem' that caused the problem, or the 'Problems of the Problem'. Hoppe (2011) argued that political struggles and political theory tended to focus solely on the tip of the iceberg (see Fig. 1.2), neglecting what lay beneath the surface.

When framing these scenarios within the context of WEMS, it should first be noted that the technological capability of WEMS as a decision-support system was never questioned; rather, it was portrayed as a tool to monitor compliance with other MEAs (see Chandran et al., 2011). As evidenced in the latter part of this book, the real problems within WEMS were intrinsic to the involvement of actors and the information they collected. The context of the objection to and redesign of WEMS

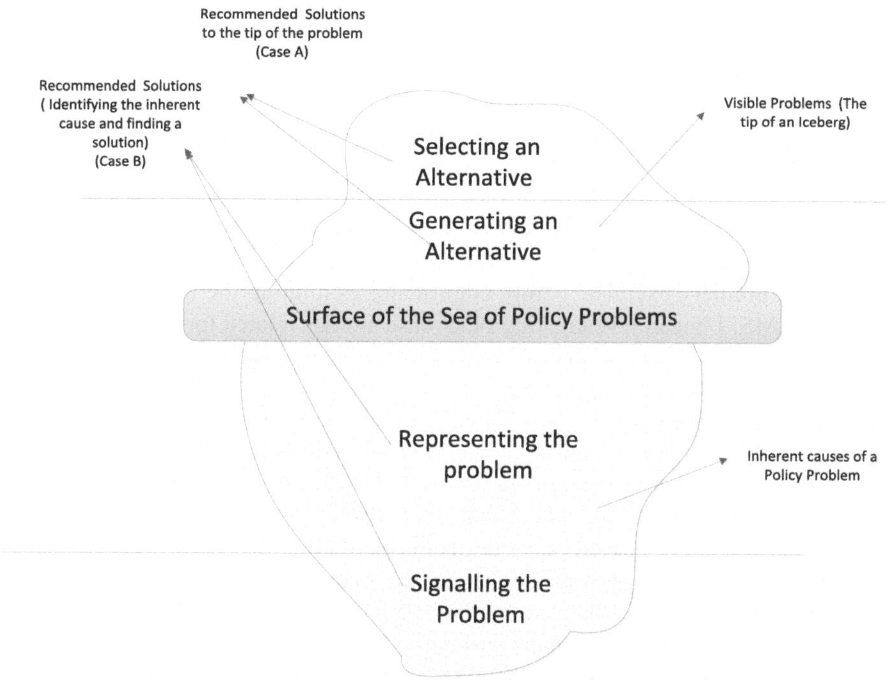

Fig. 1.2 Problem Solving and Decision-making—the Iceberg Phenomenon. The above figure describes two types of factors influencing policymaking—the visible (tip of the iceberg) and the invisible factors (hidden deep within the iceberg). Bringing out a solution to a policy problem requires addressing both the visible and invisible factors. Addressing the invisible factors requires them to be brought above the surface of the sea. The diagram is a modified version of the model proposed by Chisholm (1995)

was therefore more related to the perceptions of the actors and issues surrounding the use of information. To find a solution to the challenges within WEMS, it was essential to first identify the (hidden or inherent) factors that caused the objections. In other words, the hidden problems needed to be exposed and brought to the surface to advocate for the recommended solutions (see Fig. 1.2).

Based on the analysis of the literatures on IS and PAPPS, in this book, I use the PAPPS perspective, exploring both the visible and the more inherent or invisible factors that led to the objection of WEMS in Asia and its adoption in Africa. The visible factors can be observed from the historic proceedings of MEAs and the CITES Convention (see Chap. 2). With regard to WEMS, these visible parts are its outputs, its adoption and objection by different stakeholders (Chap. 2). The invisible factors are the perceptions of the actors working within the Wildlife–Trade policy subsystems and the external factors that influenced them (see Chaps. 4 and 5). In the next section, the research problem is defined.

1.3 Examining the Problem Within WEMS

As previously noted, the challenges faced by WEMS were largely rooted in intrinsic factors, particularly the perceptions held by various actors regarding their roles in the collection and utilisation of information. These underlying issues were often subtle and not easily identifiable. Consequently, *Silos to Synergy* delves into how individuals within the policy subsystem interpreted and responded to a specific policy issue.

As a preliminary exercise, before empirically examining the inherent problems within WEMS, I began investigating other case studies related to ICT implementations, which had also been shaped by various political dynamics. As noted in the previous section, the development of policy maps in Europe provided a classic example, demonstrating that the challenges within a policy context often extend beyond what technology can resolve. Georgiadou (2008) described problems similar to those faced by WEMS in the case of the European Spatial Development Perspective (ESDP) process, where maps visualising environmental pollution and regional development in Europe were dismissed by national policymakers during discussions on spatial issues and policies. The maps were so sensitive that it was decided not to use any of them in the final ESDP report. Similar to the challenges surrounding the sensitivity of maps during the ESDP process, the spatial display of the transboundary movement of illegal trade through WEMS also proved contentious for CITES Parties. Both examples highlighted that the use of information, whether spatial or non-spatial, was intricately tied to policy implementation concerns within specific domains, where agreement among multiple stakeholders was crucial. It should be noted that, until the policy maps were tabled for discussion, they were seen as a visual 'solution' for national policymakers. However, once the maps were scrutinised, the inherent problems surfaced, leading to their rejection.

Yet, little is known about how spatial information is used and interpreted in policy implementation (see Carlton, 2007; Dühr, 2007; Faloudi & Waterhout, 2002).

Another aspect of WEMS that required examination was the impact of the information it contained on stakeholders, including enforcement agencies, CITES, national governments, and NGOs. For instance, beyond the spatial data on the origin and destination of illegal wildlife trade, WEMS also held the names and details of criminals (nominal data). This brought WEMS to the attention of two distinct audiences: the ground-level enforcement bodies (such as police, customs, and forestry institutions) and the national policymakers responsible for wildlife management and international trade policies. While addressing wildlife crime indeed necessitated an integrated approach, bringing all these agencies together in a common decision-making forum, the challenge lay in the fact that policy problems persisted. Issues such as inter-agency coordination challenges, a Party's reluctance to acknowledge itself as a destination country for illegal trade, and the impact of funding needs on specific actors or organisations complicated matters. In reality, satisfying all the criteria of different stakeholders was nearly impossible. Nonetheless, there were likely one or two areas of 'common ground' where stakeholders could reach an agreement. This common ground could thus serve as a potential solution for WEMS adoption.

In the search for a solution to facilitate the adoption of WEMS as a multi-stakeholder information-sharing tool, my PhD supervisors at the University of Twente and the UNU team, along with myself, explored the possibility of finding a policy consensus to bring the different stakeholders (actors or coalitions) together. From the perspective of policy science, Hoppe (2011; see also Hisschemöller et al., 2001) distinguished two types of policy problems relevant to WEMS adoption: 'structured' and 'unstructured' problems. A problem was considered 'structured' or well-defined if it could be solved using standardised (quantitative) techniques and procedures. In such cases, the relevant disciplines and specialisations were clearly defined, and policymaking rested in the hands of one actor or a group of actors who agreed on the goals and means of addressing the problem. In contrast, when a problem was complex, 'unstructured' or ill-defined (often described as wicked or messy), technical methods alone were insufficient in the absence of a consensus on the goals.

Hisschemöller and Hoppe (1995) noted that governmental policymakers often preferred to define problems as 'structured', as these were easier to address. This was precisely what the UNU team attempted in the early stages of WEMS development when they tried to frame enforcement information-sharing problems as technical or structured, so that the solution prescribed would be a technical (structured) one, using the WEMS system. However, as mentioned earlier, the policymaking process for wildlife enforcement information-sharing was not within the control of a single actor. It involved multiple actors—governments, UN agencies, industry, academia, and civil society—each with their own norms and values, while also participating as multidimensional actors or coalitions in the wildlife and trade policy subsystem. When considered within this context, WEMS implementation in relation to CITES objectives was an unstructured problem, where the design preferences of one actor might not align with the logic or interests of others. In other words, to identify a solution, the issues concerning the implementation and adoption of WEMS needed to be structured in a way that facilitated its functioning or acceptance at national, regional, and international levels. WEMS should act as a thread that binds—referred to as a 'boundary object' in this book—the beliefs and norms of various coalitions within the wildlife policy subsystem.

As the above statement could be seen as a hypothesis, conducting a study to better understand the problem necessitated a focused research plan. This involved analysing the various policy problems and the actors involved within the wildlife management and CITES policy process. With this objective in mind, the next section will describe the research focus of this book, outlining the overarching research question for the study.

1.4 Thematic Focus

What can be assumed from the previous section is that the challenges encountered in the adoption and implementation of WEMS were not technical or solely IS-related, but rather policy-oriented or PAPPS-related. Secondly, within a policy context, the issue (sharing information on illegal trade) that WEMS aimed to address was complex, with a conflict of interests remaining among the different stakeholders regarding data-sharing and infrastructure management. Thirdly, few studies (such as Reeve, 2002; Wellsmith, 2011; and Phelps et al., 2010) have explored the policy-related challenges within enforcement information-sharing on wildlife crime, despite it being recognised by the UN as a pressing global issue that affects global biodiversity.

In this book, using WEMS as a case study, I evaluate the political, technical and policy challenges WEMS faced, from its early development to its eventual adoption by the Lusaka Agreement Task Force (LATF) in Africa. Based on the findings of this evaluation, I then undertake a practical exercise to propose a potential solution for addressing the problem.

The final chapter of the book summarises the transformation of WEMS into a Digital Public Infrastructure tool, thereby gaining the acceptance of a broader audience and, by generalising the issue of information to human–nature conflicts, reducing the political and policy problems associated with the system.

The specific angle taken in this book is to use a PAPPS perspective to shed light on the issues within the wildlife policy subsystem and highlight the macro-contextual effects on WEMS implementation. This is facilitated through the use of the Advocacy Coalition Framework (ACF) as the theoretical framework and Q methodology as the methodological framework for extracting policy beliefs related to second- and third-level issues. Based on the theoretical framework outlined in Chap. 3, Chap. 4 analyses the wildlife policy subsystems in Asia and Africa—focusing primarily on the beliefs and resources of the actors. Chap. 5 presents a comparative analysis of the two cases, showing how interactions within the policy subsystems influenced coalition behaviour and subsequent decision-making regarding WEMS. In Chap. 6, using Boundary Work Theory as a practical application, I propose a way forward for the global adoption and implementation of WEMS. Chap. 7 examines the transformation of a WEMS as a Digital Public Infrastructure tool to address human–nature conflicts, moving beyond the narrow concept of enforcement monitoring.

This book isolates issues of system design and use, as well as the micro-contextual effects on information system implementation.

Chapter 2
The Evolution of CITES Enforcement and the Conceptualisation of WEMS

2.1 Introduction

The conservation of global wildlife faces a profound challenge: the fragmentation of information across borders. For decades, the lack of effective information-sharing mechanisms has crippled enforcement efforts, allowing illegal wildlife trade and poaching to flourish unchecked. This chapter takes a critical look at how international systems have evolved—or failed to evolve—in addressing this issue, with particular emphasis on the role of Multilateral Environmental Agreements (MEAs) in wildlife law enforcement.

At the heart of this examination is the CITES Convention, which has long served as a cornerstone in global wildlife protection. However, the challenges surrounding its enforcement mechanisms are deeply intertwined with a complex web of political, economic and institutional barriers. Through an exploration of key turning points in the development of enforcement frameworks, this chapter sheds light on how CITES—and, in particular, its compliance mechanism—has grappled with these barriers over time.

The story of the Wildlife Enforcement Monitoring System (WEMS) serves as a case study of these challenges. The development of WEMS, a tool designed to address enforcement data silos, initially met with resistance, particularly from CITES itself, which raised concerns over its design. This chapter uncovers possible reasons or motivations behind these objections, the political and institutional factors that shaped the design of the system, and the role of stakeholders—including NGOs. It also tracks the key historical developments on enforcement information sharing and information system development.

Through this investigation, the chapter not only traces the evolution of enforcement and compliance within CITES but also highlights the systemic issues that hindered progress and the innovative solutions that are beginning to bridge these gaps. In doing so, it offers crucial insights into the future of wildlife law enforcement information-sharing, illustrating how overcoming the fragmentation of information and fostering global collaboration can reshape the enforcement landscape.

R. Chandran, *The Wildlife Enforcement Monitoring System*, Professional Practice in Governance and Public Organizations,
https://doi.org/10.1007/978-3-031-97084-9_2

By looking closely at these historical shifts and the complex forces at play, this chapter addresses the following questions:

1. What were the causes and factors for the failure of enforcement information-sharing in CITES?
2. How does the historical overview of CITES enforcement and compliance matters relate to the development of WEMS in Asia; its objection by CITES and, adoption of WEMS by the Lusaka Agreement Task Force?

2.2 Spiralling MEAs and Challenges in Enforcement Information-Sharing

Multilateral Environmental Agreements (MEAs),[1] also known as protocols, treaties and conventions, are legally binding agreements signed by countries (referred to as Parties) to address concerns related to the global environment. One key element of any MEA is the provision for information gathering and sharing about Party activities related to the agreement. Due to overarching sovereignty concerns, MEAs rely largely on self-reporting by Parties to encourage compliance to substantive control provisions (UNEP, 2007). For most MEAs, the Conference of Parties (COP) constitutes the governing body with the authority for policymaking during regular meetings. Generally, there is no clearly binding means of enforcement to MEAs, while compliance usually refers to the obligation of a Party to submit national communications and to report on key indicators (ibid).

Though the concept of MEAs existed prior to the establishment of the United Nations, its importance as a formal mechanism for transboundary environmental governance and its subsequent adoption increased dramatically soon after World War II (WWII) and the decolonisation of African and Asian countries. For instance, the pre-WWII rate of an agreement every two years increased to seven agreements per year from 1946 to 1972—the year of the United Nations Conference on the Human Environment (UNCHE) (see Mitchell, 2003). Ever since, the rate has continued to increase (see Fig. 2.1). As many as 319 agreements were completed in the 20 years between 1972 Stockholm Convention (UNCHE) and the 1992 United Nations Conference on Environment and Development (UNCED) (at the rate of an average 16 agreements per year), and 189 agreements were completed from 1993

[1] Agreements may take different forms, such as 'convention', 'treaty', 'agreement', 'charter', 'final act', 'pact', 'accord', 'covenant', 'protocol' or 'constitution' (for an international organisation) (see UNEP report, *Auditing the Implementation of Multilateral Environmental Agreements (MEAs): A Primer for Auditors*). The 1969 Vienna Convention on the Law of Treaties defines a 'treaty' as 'an international agreement concluded between States in written form and governed by international law, whether embodied in a single instrument or in two or more related instruments and whatever its particular designation' (See UNEP report, *Auditing the Implementation of Multilateral Environmental Agreements (MEAs): A Primer for Auditors*). In this book, as a practical matter, 'treaty', 'convention' and 'agreement' are often used interchangeably.

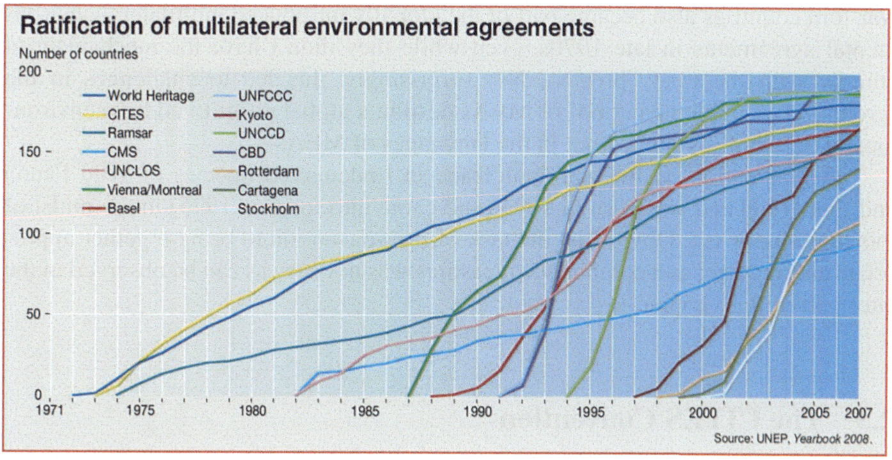

Fig. 2.1 Ratification years of multilateral environmental agreements and the number of Parties (source UNEP, 2008)

through 2002 (19 agreements per year) (Mitchell, 2003). Almost half of all MEAs attempt to protect species or manage human impacts on those species.

As all MEAs outline strict sovereign rights[2] to the states or Parties; with the increase of MEAs came the responsibility for individual Parties to report on the status of the environment within their jurisdiction. But challenges remained in implementation and reporting as the timing of the formation of most MEAs coincided with the recovery from WWII and the decolonisation processes. Apparently, the parties were not prepared to meet the requirements, leading to weak compliance with MEAs. For instance, in many developing countries, spatial and non-spatial data on the general state of affairs of the global environment on biodiversity is still not available at a national level. For these countries, scientific institutions in the developed world remain the only source of credible information. In such circumstances, Parties compromise their sovereignty by default, partly due to their inability to carry out the reporting exercise or partly because of their historical ties with the former colonial powers in mapping and supporting research on natural resources. This particular characteristic of science and reporting, where information on biodiversity distribution from the developing world is tied with the works of actors in the developed world, should be taken into consideration while attempting to study self-reporting by Parties.

The implementation of the *Declaration on the Granting of Independence of Colonial Countries and Peoples* (also known as the Special Committee on Decolonization, or C-24) in 1961, led to independence of several territories in Africa and Asia which were earlier part of the western colonial regimes. But the independence soon led to power struggles and civil wars, especially in many parts of Africa and Asia that were rich in natural resources. These newly formed resource-rich but

[2] Meaning that only states and respective agencies can be the custodians of 'common pool resources', such as plants and animals, within their jurisdiction.

war-torn countries also became part of the globally introduced multilateral environmental agreements in late 1970s, even while they didn't have the mechanism to comply with the treaty process. Not surprisingly, this led to challenges in the 'ground-level implementation' of MEAs despite a global effort to address environmental problems as spelled out in the languages of MEAs.

The Convention on International Trade in Endangered Species of Wild Fauna and Flora (CITES) was born in 1973 (and came into force in 1975) in the midst of this tumultuous era. Obviously, the task of the convention to bring together appropriate enforcement and compliance measures was not easy as can be observed in the later part of this chapter.

2.3 The CITES Convention

CITES is considered as the flagship global wildlife agreement, which was formed at a time when the environmental movement was at its peak and when wildlife trade was largely one of the niche demands for wealthy customers of former colonial powers. Apparently, the conservation movement also started from western countries where the concern for trade and exploitation of wildlife in its former colonies created a strong urge among its conservation NGOs to pressure their governments to protect wildlife, mainly in its former colonies.

The convention text was originally drafted as a result of a resolution adopted in 1963 at a meeting of members of IUCN (International Union for the Conservation of Nature, commonly referred to as the World Conservation Union). The text was agreed upon at a meeting of representatives of 80 countries, in Washington, D.C., on third March 1973, and on first July 1975, CITES entered into force (CITES, 2012d). As of 2013, 178 member states (referred to as Parties) are signatory to the convention (see Fig. 2.2). The primary objective of CITES is to ensure that international trade in specimens of wild animals and plants does not threaten their survival.

Since its formation, CITES has been trying to ensure full compliance by its Parties. Through resolutions and decisions of the Conference of the Parties (COPs)—or the so-called soft laws—CITES Parties are obliged to align national laws with the regulations of the convention. In the 1979 CITES COP meeting, CITES established the Standing Committee (SC), and later the Animals and Plants Committees, Identification Manual Committee and Nomenclature Committees. These committees were given permanent status in 1987 (IUCN, 2000). Implementation of the international law of CITES also requires national governments to bring their enforcement units (Police, Customs, Military and Forest officials—the only type of agencies tasked with actual implementation) to identify and confiscate illegal wildlife specimens which are listed in the convention as potentially threatened.

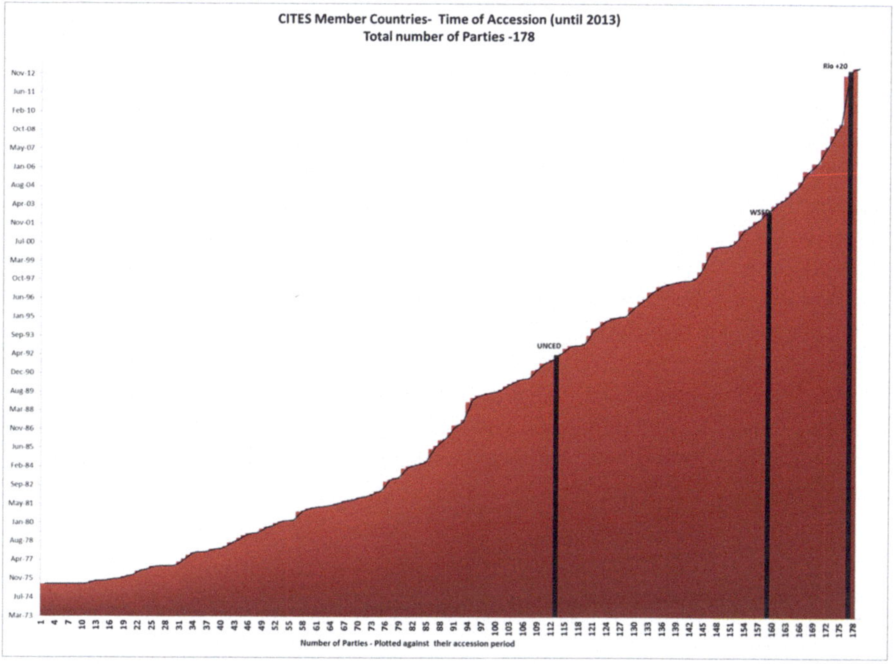

Fig. 2.2 CITES member countries and their time of accession (1973–2013) (CITES, 2011c). The bold vertical lines indicate the timeline of the UN Conference on Environment and Development

The convention text, especially Articles VIII[3] and XIII,[4] emphasise the need for Parties to maintain records of wildlife trade and transmit such information through

[3] Article VIII

6. Each Party shall maintain records of trade in specimens of species included in Appendices I, II and III which shall cover:

(a) The names and addresses of exporters and importers and.
(b) The number and type of permits and certificates granted; the States with which such trade occurred; the numbers or quantities and types of specimens, names of species as included in Appendices I, II and III and, where applicable, the size and sex of the specimens in question.

7. Each Party shall prepare periodic reports on its implementation of the present Convention and shall transmit to the Secretariat:

(a) An annual report containing a summary of the information specified in sub-paragraph (b) of paragraph 6 of this Article and.
(b) A biennial report on the legislative, regulatory and administrative measures taken to enforce the provisions of the present Convention.

[4] *Article XIII* -International Measures

1. When the Secretariat, in the light of information received, is satisfied that any species included in Appendix I or II is being affected adversely by trade in specimens of that species or that the provisions of the present Convention are not being effectively implemented, it shall communicate such information to the authorised Management Authority of the Party or Parties concerned.

annual and biannual reports. Article XIII defines the steps the CITES Secretariat should take to address any issues regarding any adverse trade that affect the survival of the species. Failure to implement the regulatory provisions of CITES, or to report instances of non-compliance, may result in sanctions if endorsed by the Standing Committee and the Conference of the Parties (CoP) to the Convention. However, CITES regulations do not supersede national legislation; their primary purpose is to assist the Parties in implementing the Convention, with the responsibility for implementation resting solely with the Parties themselves (CITES, 2013b).

In addition, CITES receives substantial financial support from the European Commission and the United States. Other major donors include Denmark, France, Germany, Hong Kong SAR (a Special Administrative Region of China), Japan, Norway, Qatar, Sweden, and the United Kingdom of Great Britain and Northern Ireland. These countries continue to provide funding for activities such as capacity building, science-related initiatives, national legislation, enforcement, the Sponsored Delegates Project, and the Monitoring the Illegal Killing of Elephants (MIKE) programme (CITES, 2013a).

2.3.1 The CITES Compliance System

The CITES 'compliance system' is defined as the 'subset of the treaty's rules and procedures that influence the compliance level of a given rule'. The level of compliance of a State is measured based on the actions by national governments within three regulatory systems (Reeve, 2001):

(a) The primary CITES rule system.
(b) The compliance information system.
(c) The non-compliance response system.

The chief regulating bodies within the CITES compliance system are the COPs, the Standing Committee and the CITES Secretariat. The COP is composed of Party representatives and is the supreme decision-making body, meeting every two and a half years. The Standing Committee is an executive body made up primarily of 14 representatives of the Parties. It is elected on a regional basis and oversees the operation of the convention between CITES meetings. Its functions include: overseeing financial activities; coordinating and advising other committees, as well as working groups set up by the COP; drafting potential CITES resolutions and performing any other functions as may be entrusted to it by the COP (Reeve, 2002, pp. 47).

---

2. When any Party receives a communication as indicated in paragraph 1 of this Article, it shall, as soon as possible, inform the Secretariat of any relevant facts insofar as its laws permit and, where appropriate, propose remedial action. Where the Party considers that an inquiry is desirable, such inquiry may be carried out by one or more persons expressly authorized by the Party.

The Animals and Plants Committees, composed of technical experts (usually biologists), are elected on a regional basis and play a minor role in the compliance system. Reporting to the COP and the Standing Committee (if requested), these technical bodies are responsible for reviewing the status of selected species and advise on the action to be taken. In a way, this process within the CITES compliance mechanism ensures that the decisions taken by CITES authorities rely on a science-policy mechanism where there is an interaction between national management (including policymakers and scientists at the national level) and science authorities (for example, IUCN and the World Conservation and Monitoring Centre—WCMC) at the international level. In this regard, CITES is different from other conventions, such as the United Nations Framework Convention on Climate Change (UNFCCC), where an independent science body like the Intergovernmental Panel on Climate Change (IPCC) plays a leading role in bringing science-based decisions to UNFCCC and the Conference of Parties.

Non-governmental organisations (NGOs) are also crucial participants in the CITES compliance system (see Article XII[5] of CITES Convention). The International Union for Conservation of Nature (IUCN) and TRAFFIC—the joint programme of the World Wide Fund for Nature (WWF)[6] and IUCN—play pivotal roles in identifying and flagging non-compliance by CITES Parties. Additionally, other NGOs exert considerable influence within CITES, primarily through lobbying efforts at CoP meetings.

2.3.2 Primary Rule System

The convention regulates international trade in wildlife through a permit system that is applied to species listed in three Appendices (Fig. 2.3):

- Appendix 1—Prohibiting commercial trade. Only non-commercial trade, largely for scientific and educational purposes is allowed.
- Appendix 2—Restricted trade under which commercial trade is controlled.
- Appendix 3—It includes species listed unilaterally by Parties needing international assistance to control trade.

The Appendices are revised at each CITES meeting where a two-thirds majority is required for amendments to be adopted. As CITES only regulates trade and does not prevent it (as seen from the large volume of species in Appendix II, Fig. 2.3), decisions for species to be listed becomes political during a time of no consensus

[5] Article XII Upon entry into force of the present Convention, a Secretariat shall be provided by the Executive Director of the United Nations Environment Program. To the extent and in the manner he considers appropriate, he may be assisted by suitable inter-governmental or non-governmental international or national agencies and bodies technically qualified in protection, conservation and management of wild fauna and flora.

[6] Also known as World Wildlife Fund

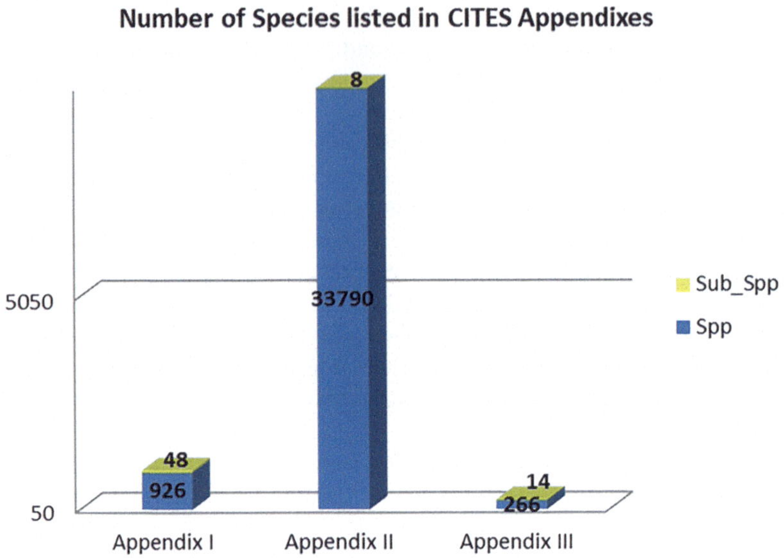

Fig. 2.3 Total number of species (Spp.) and sub-species (*Sub*-Spp.) (References related to species and subspecies are available at the CITES website: http://www.cites.org/eng/disc/species.php) listed in CITES Appendices as of 2013

(for example, decisions regarding permitting ivory trade). All trade in listed species must have a permit or certificate. The listing of species is increasing, and, as of 2025, over 40,900 species—including roughly 6610 species of animals and 34,310 species of plants are listed in Appendix 1, II and III (See Fig. 2.4).

As mentioned earlier, CITES is a non-self-executing treaty, meaning that national legislation is required to implement several provisions. Parties are required to designate 'Management Authorities' and 'Scientific Authorities' competent to grant permits or certificates. The 'Scientific Authorities' play an important role in verification through monitoring exports permits and producing non-detriment findings (Reeve, 2001). Although such a procedure is mandatory for maintaining compliance, Management and Scientific Authorities in several developing countries often remain under-resourced, understaffed and, in some cases, non-existent, as will be seen in later sections of this chapter.

2.3.3 *Compliance Information System*

CITES is one of the first MEAs to develop an information system to monitor enforcement of and compliance to the convention. The collection, review and dissemination of enforcement data are the responsibilities of the Secretariat. Though CITES largely relies on self-reporting by Parties, it also seeks information from NGOs and relevant intergovernmental organisations. In parallel, information is also

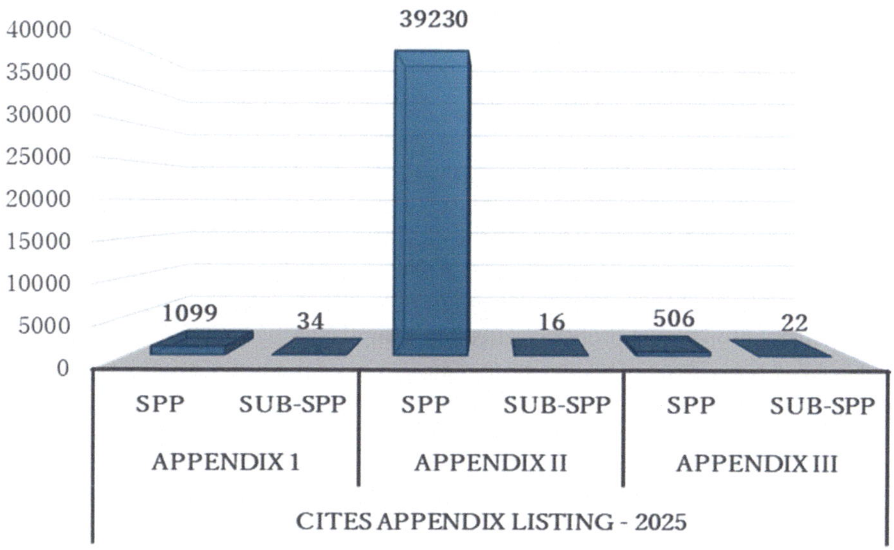

Fig. 2.4 Total number of species (Spp.) and sub-species (*Sub*-Spp.) listed in CITES Appendices as of 2025 (Updated list of CITES Species listed as of seventh February 2025 https://cites.org/eng/app/appendices.php)

gathered by the Secretariat during ad-hoc visits to Parties, usually at the request of the COP or the Standing Committee (Reeve, 2002).

As a CITES rule or compliance mechanism, Parties are required to provide enforcement and compliance information (how the convention is enforced into the national laws, and the level of compliance with CITES provisions) through annual and biennial reports. Annual reports include information on trade in CITES specimens, while biennial reports contain data on legislative, regulatory and administrative steps taken to enforce the convention. While several Parties do comply with the annual reporting process, biennial reporting has remained largely unimplemented (Reeve, 2001). This has led CITES to merge the existing reporting format into one national report (CITES, 2013d) and one implementation report so as to bring down the burden of the reporting process.

Upon receiving the report, the Secretariat is mandated to study Parties' reports to request further information and to prepare annual reports on implementation. Due to lack of manpower and expertise at the Secretariat to carry out all the functions, some of these functions are contracted out. For example, the trade records from Parties are maintained in a database by UNEP-World Conservation and Monitoring Centre (WCMC) which has been in existence since 1975. WCMC is also tasked with comparing the import-export records and, where the records do not match, or Parties report possible illegal trade, the WCMC informs the Secretariat of cases of non-compliance. Annual reports are essential for analysing trade in CITES-listed species. Trade studies are heavily dependent on precise and complete reporting by Parties. Yet, complete reporting has turned to be a persistent problem primarily because several Parties fail to comply with the timely reporting responsibilities (CITES, 2012c; Phelps et al., 2010).

Until CITES COP 11 (April 2000), the Secretariat had been compiling comprehensive data on illegal trade in a Report on 'Alleged Infractions' received from Parties, NGOs and other sources, such as INTERPOL and the WCO. Using this information, two types of infractions had been detailed prior to COP 11: (i) illegal trade commonly committed by individuals and (ii) non-compliance by Parties with the provisions of the convention. The infraction reports were then made public (since COP 12, only a limited number of cases were made public), providing all Parties with a record of significant violations that took place between COP meetings and of other enforcement problems affecting compliance. They were then discussed and debated for seeking mechanisms to reduce or to eliminate the problems. Despite some Parties complaining about having their violations placed on record, infraction reports came 'to be accepted as a reliable and impartial instrument reinforcing national implementation and accountability' (Reeve, 2001).

In CITES compliance process, various NGOs (including industry representatives such as captive breeders as NGOs) work closely with the CITES Secretariat, Governments and sometimes with traders, either individually or in partnerships. While IUCN-affiliated NGOs such as WWF and TRAFFIC (The Wildlife Trade Monitoring Network) play a significant and important role in assisting the CITES Secretariat in its administrative tasks, other conservation NGOs act as watchdogs to the compliance with the convention or engage in capacity development work.

The Secretariat receives information on compliance from NGOs either directly or indirectly via reports from Parties to the Secretariat. Since its establishment in 1976, TRAFFIC is one of the key NGOs that collects information on illegal wildlife trade and continues to be engaged in transmitting information to the Secretariat and national authorities. TRAFFIC, on several occasions, has assisted the Secretariat in enabling countries to comply (see sect. 2.4). TRAFFIC also maintains the Elephant Trade Information System (ETIS, CITES, 2013a), which was set up to monitor the ivory trade.

Although IUCN and its main subdivision, the Species Survival Commission (SSC), do not have any legally binding authority, their opinions are considered dependable: governments, scientists, journalists and others use it as a quick, reliable way to find out whether particular species are in danger of extinction, without having to undertake lengthy research (Rodrigues et al., 2006). In spite of their larger role in building a science-based decision-making process, some scholars remain sceptical of the role of IUCN in influencing CITES decisions and on how they use their privileged positions to influence CITES COP outcomes (Mrosovsky, 1997).

2.3.4 Non-Compliance Response

Until 2010, the CITES Secretariat employed just one senior enforcement officer to police around 35,000 animal and plant species listed in its appendices. As a means of verification, the CITES Secretariat conducts ad hoc visits or missions to Parties experiencing implementation problems, and these missions are mainly based on the approval of the Standing Committee. The purpose is to gather information, assess problems and provide advice to national authorities.

In sum, the CITES compliance mechanism does have an overall good regulatory framework at a global level. However, in monitoring the enforcement of the convention at a national level, especially in countries where governance mechanisms do not function or do not exist, the task of compliance monitoring remains highly challenging. To understand the challenges in the monitoring process of the convention, the next section explores the historical timeline of the events that took place and which shaped the enforcement information processes within the convention.

2.4 CITES Enforcement: A Timeline of Information Sharing Efforts

As mentioned in Sect. 2.3, The CITES Convention was established at a time when the need to protect the environment was a shared concern among numerous governments. While the convention was founded with great optimism, securing financial mechanisms to implement it remained one of its most significant challenges. To address this, during the second CITES Conference of the Parties (CoP), UNEP's Executive Director, Mustafa Tolba, echoing the sentiments of his predecessor, Maurice Strong, concluded that the most effective way for UNEP to fulfil its responsibility to CITES was through an agreement with the IUCN, under which the IUCN would provide the staff and resources to carry out the Secretariat's functions. (CITES, 1979). This marked the beginning of the long-standing cooperation between IUCN and its affiliates (WWF, TRAFFIC, IUCN Environmental Law Programme) and the CITES Convention, during which IUCN, for a time, shared office space and staff with the Secretariat. Indeed, some of the staff working for the CITES Secretariat were seconded from IUCN. The decision to involve IUCN was based on its extensive expertise and long-standing commitment to the conservation of wildlife, as well as its pivotal role in the preparation for the Plenipotentiary Conference in Washington, which ultimately led to the formation of the Convention. (CITES, 1979).

The late 70 s and early 80 s were also crucial with regard to membership of the Parties to the Convention. For example, during the first COP meeting of the convention held in 1976 in Bern, Parties adopted their first set of criteria, referred to as the Bern Criteria. According to some researchers (Young, 2003), the Bern Criteria placed strong preservationist (or protectionist) values in the listing of species where the decisions were based heavily on the species' biological and trade status—a criteria used to establish the standards for Appendix I and Appendix II listings (Young, 2003). At the time, East Asian countries were not parties to the CITES Convention. However, when Indonesia became a Party to CITES in 1979, three major East Asian nations—Japan (1980), China (1981) and Thailand (1983), all of which had significant trade relations with Indonesia—subsequently joined. These three countries would later become a powerful political force in the broader decision-making processes within the Convention. One of the key turning points came with the shift from the Bern Criteria to the adoption of the Fort Lauderdale Criteria at the Ninth Conference of the Parties in 1994 (Resolution Conf. 9.24 Criteria, 1994). This change marked a shift from a preservationist approach to a more utilitarian one at CITES CoPs (Young, 2003). However, the utilitarian concept was not explicitly articulated at CITES meetings. Instead, it was veiled under the scientific term 'Sustainable Use' (see Chap. 3).

Until the late 1980s, CITES was also one of the MEAs which maintained a transparent approach in its proceedings. Illegal trade or infractions were discussed openly during the Conference of Parties. But with more Parties joining the convention and increased trade taking place, the reporting of infractions suffered several challenges. A crucial incident that led to the closing of the lid on CITES enforcement information openness can be traced back to the sixth CITES Conference of Parties held in July 1987 in Ottawa. In documents made public in Ottawa, the CITES Secretariat named 40 member countries for suspected violations, and asked them for written explanations (CITES, 1987). The officials noted that they could easily have listed many more. Among the Parties, three major wildlife importing countries—France, Japan and Austria—came in for special criticism from several delegations and private groups, including NGOs. Critics charged France for being lax in controlling imports and for allowing nearly unfettered trade in illegal products through French Guiana, a French department in South America. Japan was criticised for importing large volumes of illegal products and for exempting itself, as it legally can under the treaty, from prohibitions on the import of shells of endangered sea turtles, glands of musk deer and several other items. Austria was criticised for allowing a pharmaceutical company to import 20 chimpanzees from Africa, which the treaty staff asserted was against the rules. France denied the charges, the Japanese said they were working to improve compliance, and Austria argued that the chimpanzee shipment had been legal (Eckholm, 1987). While all three expressed bitter resentments at being singled out, the then Director of TRAFFIC international, Ginette Hemley in an interview with *New York Times* said, '…the frankness about treaty infractions is an encouraging development…' (Eckholm, 1987) (Fig. 2.5).

The challenges in enforcement were partly due to the fact that many new member states (especially from developing countries) did not have a clear mechanism to bring its enforcement agencies to comply with CITES. National customs, police and other enforcement units largely remained unclear of what the provisions of CITES entailed, and the identification of specimens was simply not an easy task. The agencies till then had never before worked together on controlling wildlife products, and especially the Customs agencies were new to the provisions of CITES rules and regulations.

With several countries continuing to show poor enforcement, and NGOs reporting the failure to the CITES Secretariat, the conflict between the sanction-inflicted countries and international environmental NGOs became more serious.

In October 1989, during the seventh COP of the convention held at Lausanne, Parties collectively expressed their concern and disillusionment about their name

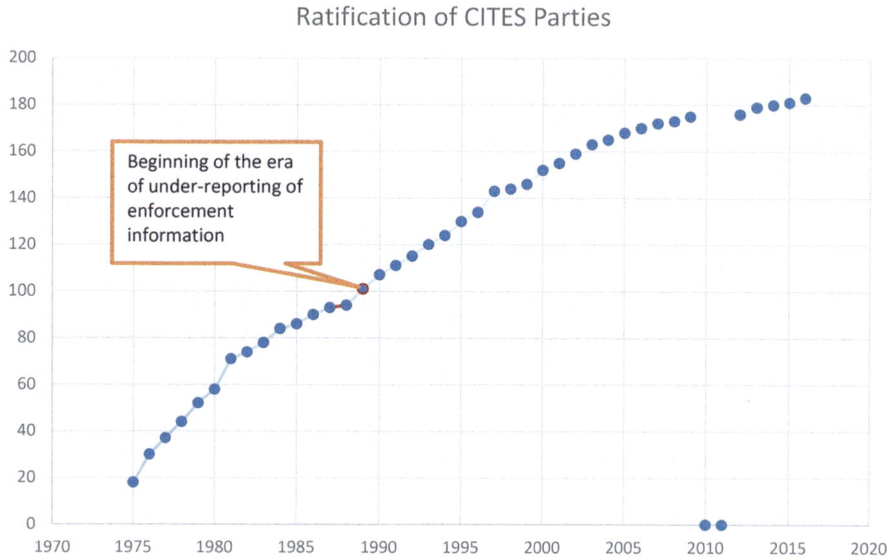

Fig. 2.5 Year of ratification of Parties, indicating the period of underreporting

being mentioned (CITES, 1989). The national CITES Management authorities attending the meeting said it was not to blame for the non-compliance process which took place beyond their official control. The Secretariat however, stressed that it is the State which constitutes the Party to the Convention which requires all State services to work towards the application of the Convention. But this was not convincing for the Parties and, as it turned out, many Parties did not respond at all or, only rarely so to the requests for information made by the Secretariat on the basis of Article XIII of the convention. Some others replied so late that their response was no longer useful in resolving the problem at hand.

Another crucial factor that clogged the enforcement information-sharing and the openness of CITES decision was the African Elephant issue. With rampant poaching of elephants in Africa in the 1980s, the United States' approach to trade in ivory changed. In 1989, President George H.W. Bush unilaterally banned ivory imports. Kenya burned its 13 tons of ivory stocks, and CITES announced a global ivory ban, which began in 1990. But not all countries agreed to the ban. Zimbabwe, Botswana, Namibia, Zambia and Malawi expressed 'reservations', exempting them from it on the grounds that their elephant populations were healthy enough to support trade. Blamed with serious implementation challenges and with non-compliance steadily increasing among the Parties, CITES was forced to resort to sanctioning the non-compliant states.

In 1992, Italy was the first European country to receive trade suspension by CITES based on a report filed by TRAFFIC (Reeve, 2001). During the 1990s, several African countries also went through trade sanctions even though they hardly had any mechanism to implement the CITES provisions. As it is clear now, enforcement will be effective only if there are sufficient resources for monitoring and managing the national park service (Hilborn et al., 2006). At that time, many developing nations lacked both the requisite legislation and the necessary resources. While civil society advocates in the United States, the United Kingdom and France were able to lobby their respective governments for internal action against illegal trade, the absence of democratic opposition to environmental crime in certain Asian countries allowed for widespread violations of the convention. This situation significantly undermined the implementation and efficacy of the convention itself, contributing to the insular approach taken by the CITES Secretariat in handling enforcement-related information (see Fig. 2.4).

Almost two decades after the formation of CITES, the compliance mechanism of the convention was getting immensely complicated. The eighth meeting of the CITES COP (held in Kyoto, Japan, in 1992) opened with the CITES Secretariat reiterating the need for submission of annual reports and highlighting that non-submission itself was a form of infraction (CITES, 1992). During that period, the CITES Secretariat, comprising a modest staff of just 30, was tasked with overseeing compliance across 115 countries and regulating a wide array of species listed in its appendices. Faced with acute resource constraints—and with wildlife protection ranking low on the policy agenda of many states—the Secretariat's responsibilities became exceptionally burdensome. It seemed that the only practical option was to delegate certain functions to trusted non-governmental organisations and consultants. It should be noted that while most conservation NGOs engage in some form of collaboration, certain ideological[7] differences remain—particularly concerning the application of the precautionary principle. Within the context of CITES, this principle supports the adoption of trade restrictions or protective measures even when scientific evidence of harm is not yet conclusive, provided there is a reasonable suspicion of risk. In essence, the absence of complete scientific certainty should not justify delaying action to safeguard a species. This position contrasts with the more conventional, science-led approach that favours waiting for robust evidence before intervening, reflecting a broader divide between precautionary ecological thinking and traditional modernist perspectives (Fig. 2.6).

While the struggle between trade and conservation was complicating the decision-making process of CITES, several parallel programmes by the United Nations also affected the funding and compliance mechanism (due to the overlapping and often conflicting mandates with other MEAs) of the convention. Some of the events that occurred during 1990s, including the formation of the World Trade Organisation (WTO) in 1992 and the Convention of Biodiversity (CBD) in 1993, took away a lot of attention from the CITES convention. The Parties that were

[7] IUCN considers itself as a neutral forum for governments, NGOs, scientists, business and local communities to find practical solutions to conservation and development challenges. While the conservation objectives of IUCN are broad, the interests of some individual members may be much more concerned with protection and others with utilization, and based on circumstances, they take individual positions accordingly. The Species Survival Network believes in a more protectionist position and that trade can occur only when the evidence positively demonstrates that the survival of the species and their role in the ecosystems in which they occur will not be detrimentally affected by trade. The species must always receive the benefit of the doubt if the available evidence is uncertain.

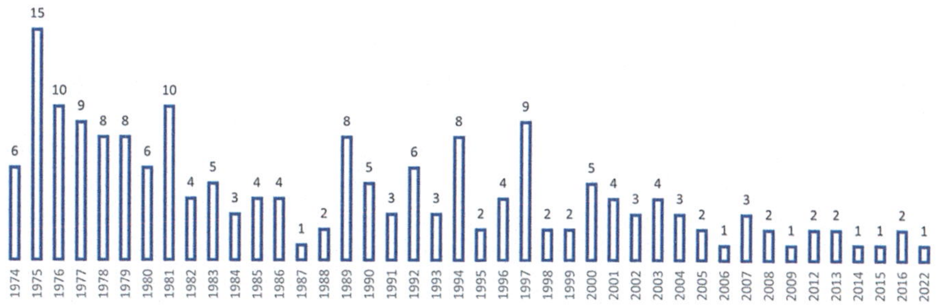

Fig. 2.6 The number of parties joining CITES Convention each year

signatory to CBD and WTO were also part of the CITES process and this at times conflicted each other's mandate. Several other MEAs also sprouted during the 1990s, imposing severe administrative processes for Parties to achieve national compliance with MEAs in general. The impact of all these developments became more evident in the later events of the convention and especially during the CITES COP 12 meeting.

Recognising the challenges of undertaking enforcement efforts in isolation, CITES proposed in 1993 the establishment of a Law Enforcement Network comprising enforcement officers nominated by the Parties. However, at the 31st meeting of the Standing Committee in March 1994, the Secretariat reported mixed reactions from Parties regarding the creation of a new, standalone mechanism within CITES to address enforcement matters. The Committee ultimately decided to rely on existing intergovernmental enforcement mechanisms, rather than establishing a separate structure within the Convention. (CITES, 1994b, 1995).

The purpose of this form was to enhance communication between INTERPOL's National Central Bureaus (NCBs) and the General Secretariat on crimes involving the international trade and transport of dangerous goods, nuclear waste, and wild fauna and flora. In support of this initiative, the CITES Secretariat issued a notification to Parties, requesting their wildlife management authorities to submit relevant information using the ECO-MESSAGE[8] format to their respective NCB (CITES, 1995) who can then send such information to INTERPOL.

With the focus turning to working with existing intergovernmental enforcement mechanisms, the Secretariat agreed to increase its cooperation with ICPO-INTERPOL and the World Customs Organisation (WCO) (CITES, 1997a). The Secretariat signed a memorandum of understanding (MoU) with INTERPOL on 2 January 1996 and with the WCO on 4 July 1996 (CITES, 1997a, 1997b). The main objectives of this MoU were to:

- Exchange information and intelligence on wildlife crime.
- Cooperate in the training of police and customs officers.

[8] The ECO-MESSAGE form went through a modification process in 2008 to make it convenient for Parties to use (CITES, 2009).

There was also an idea of establishing a common intelligence database between the Secretariats of World Customs Organisation (WCO) and CITES (with a possibility of INTERPOL joining, CITES, 1997a, 1997b). However, due to insufficient budgets and the lack of manpower, such a system never came into existence.

With traditional enforcement agencies partnering with the CITES Secretariat, the same principles governing enforcement information also applied to the databases subsequently developed by these agencies. When INTERPOL established its Wildlife Working Group, it acknowledged that membership of the group on Illegal Wildlife Trade consisted solely of enforcement personnel from the CITES Secretariat (typically one officer) and government representatives from Parties with primary enforcement responsibilities under the Convention, ensuring representation from each geographic region. INTERPOL also sought clarification regarding the group's membership—whether it comprised Parties or individuals—and on the participation of observers. The involvement of NGOs has, ever since, been the subject of protracted discussion at each meeting of INTERPOL's sub-working group. As ECO-MESSAGEs could only be submitted to INTERPOL by a National Central Bureau (NCB), CITES Parties first needed to persuade their wildlife divisions to collaborate with national enforcement agencies. However, this inter-agency coordination proved challenging, and the submission process itself remains, to this day, ambiguous. In effect, there was little alternative but for wildlife divisions to send ECO-MESSAGEs directly to the CITES Secretariat, which in turn had to maintain its own database on illegal wildlife crime. Yet, the already overstretched Secretariat lacked both the capacity and resources to establish and sustain such a database.

A turning point came in 1997, when a protracted debate over a proposed one-off sale of ivory to Japan culminated in a consensus decision at the tenth meeting of the CITES Conference of the Parties (COP10) in Harare (CITES, 1997b). One of the preconditions for the sale was the establishment of an international reporting and monitoring system for illegal international trade, maintained through a database housed within the CITES Secretariat—later known as the Monitoring of Illegal Killing of Elephants (MIKE). Additionally, an international reporting and monitoring system was to be established for illegal trade and poaching occurring within or between Elephant Range States, managed through a database overseen by TRAFFIC International. Other bodies, such as the IUCN/SSC African Elephant Specialist Group and the Lusaka Agreement Task Force (LATF), were invited to collaborate in the development of these elephant databases. At the meeting, TRAFFIC was formally contracted to develop a comprehensive database dedicated to monitoring illegal trade in elephant ivory and its derivatives, known as the Elephant Trade Information System (ETIS).[9] Around the same time, the CITES Secretariat resolved to develop the Trade Infraction and Global Enforcement Recording System

[9] Since its inception, ETIS has been managed by TRAFFIC on behalf of the CITES Parties and is currently housed at the TRAFFIC East/Southern Africa office in Harare, Zimbabwe. ETIS is an improvised version of Bad Ivory Database System (BIDS) managed by TRAFFIC since 1989. The funding for the development of ETIS mainly came from WWF, DEFRA and the CITES Secretariat (CITES, 2000a).

(T.I.G.E.R.S.)—the first CITES database dedicated to recording seizures and infractions. (CITES, 2000b). T.I.G.E.R.S. operated as a single-channel information process where Parties could send the infraction cases in an ECO-MESSAGE format.

At CITES COP 11 held at Nairobi, several countries, including US, UK, Belgium and Japan, pledged financial support to ETIS and MIKE—the two elephant databases (CITES, 2000a).

While CITES received financial support for building information systems as a requirement for bridging the consensus sale of ivory, scepticism increased when the ivory sale coincided with an increase in elephant poaching. At COP 11 in Nairobi in April 2000, two conservation NGOs, the David Shepherd Conservation Foundation and the International Wildlife Coalition (IWC), both at that time members of the Species Survival Network (SSN), questioned the scientific validity of the CITES Secretariat's conclusions on the sale of ivory. They also claimed that such a sale had led to rampant poaching. TRAFFIC (though being a NGO) supported CITES and contested NGO figures on poaching, suggesting double counting (ENB, 2000a, 2000b).

Grappling with NGOs and the elephant issue, at COP 11, the Secretariat unilaterally decided to redefine the reporting process on infractions. CITES stated that the goal of the reports is to provide an 'overview of "illicit trade" and to identify "significant problems" relating to the issuance and acceptance of CITES documents and henceforth to report only work by Parties that illustrates innovative or particularly significant enforcement action' (CITES, 2000b). This decision resulted in a COP 11 report having just six pages of mostly general information, compared with almost 100 pages detailing infractions in previous years (Reeve, 2001). Only three infraction cases were mentioned in anything more than general terms. Prior to this decision, two types of infraction were detailed:

(i) Illegal trade, commonly committed by individuals.
(ii) Non-compliance by Parties with the provisions of the convention.

According to Reeve (2001), infraction reports were considered a reliable and impartial instrument that reinforced national implementation and accountability, and the unilateral shift by CITES was a betrayal on the term 'infractions' (See Fig. 2.4). The elephant databases and T.I.G.E.R.S., developed during this period, were accessible only to their respective custodians: TRAFFIC and the CITES Secretariat for ETIS; the CITES Secretariat and IUCN for MIKE; and solely the CITES Secretariat for T.I.G.E.R.S.

At the 12th CITES meeting held in Santiago, Chile, in November 2002, the Executive Director of UNEP highlighted the importance of the World Summit on Sustainable Development (WSSD, the timeline of WSSD indicated in Fig. 2.1) and, in particular, the Plan of Implementation of the Johannesburg Declaration (CITES, 2002b). He also noted that, in light of the Doha Development Agenda (WTO,

2001),[10] it was essential for multilateral environmental agreements (MEAs) to coordinate their activities with the World Trade Organization (WTO), given that all environmental conventions encompassed a trade dimension. UNEP's decision to align MEAs with WTO principles received strong backing from IUCN. The IUCN observer expressed firm support for the concept of sustainable use, emphasising the benefits of sustainable trade in wild species, particularly in relation to the seventh Millennium Development Goal, which focuses on the alleviation of poverty.

During CITES COP12, the Secretariat claimed that the shorter reporting form (since it stopped publicly listing complete infraction cases) had led to changes in attitude among Parties and that there was a marked increase in the supply of information by some Parties that were previously uncommunicative (CITES, 2002a). However, this claim was not entirely accurate, as evidenced by developments at COP14 (CITES, 2007b). Beyond the political sensitivities, Parties faced several practical obstacles in providing information. As previously noted, the CITES compliance process imposed a substantial administrative burden on operational staff, who were consequently reluctant to engage in additional reporting beyond their mandated responsibilities. Furthermore, data protection laws and other national regulations limited the ability of certain countries to share information concerning offenders and criminal activities through their wildlife divisions. Reporting crime was also perceived by some staff as a reflection of failure in fulfilling their enforcement duties, further discouraging transparency.

As an exception and to lighten the burden, CITES Secretariat communicated that it was ideal to have the information submitted as an ECO-MESSAGE, but that the format used was less important and that CITES would accept information communicated by whatever means were most convenient for the Parties (CITES, 2002a). Furthermore, if Parties encountered difficulties in transmitting information directly to the Secretariat, they were encouraged to utilise alternative channels provided by the World Customs Organization and ICPO-INTERPOL—both of which had formal agreements with the Secretariat to facilitate the exchange of information.

At CITES COP 12, it was getting clearer that submission of information by the Parties was slowing down. The CITES Secretariat was also conscious that the national enforcement staff was unwilling to contribute to a system that was not of benefit to them. The level of reporting to T.I.G.E.R.S. database was so low that it could not provide any meaningful analysis of trends in illicit trade (CITES, 2002a).

Providing information to the Secretariat was only one facet of compliance; tangible enforcement on the ground was also lacking, as many Parties simply did not have the capacity to implement the Convention effectively. In most countries, the responsibility for enforcement lay with wildlife divisions whose expertise did not

[10] The Doha Development Agenda or Doha Round is the round of trade negotiations among the World Trade Organization's membership. Its aim is to achieve major reform of the international trading system through the introduction of lower trade barriers and revised trade rules. The work programme covers about 20 areas of trade. The Round is also known semi-officially as the Doha Development Agenda as the fundamental objective is to improve the trading prospects of developing countries.

extend to criminal investigations. Consequently, they were ill-equipped to address illegality relating to imports, exports, and domestic sales. Amid the complexities of information-sharing, CITES even proposed that 'real-time' information exchange could be more effective than waiting for reports submitted every two and a half years. However, for such a system to succeed, it required regular feedback from Parties—something that was clearly lacking. For instance, many of the faxes the Secretariat sent to Parties requesting information on illicit trade, forged documents or reported seizures were neither acknowledged nor responded to (CITES, 2002a).

Despite severe shortages in manpower and persistent gaps in information flow, the CITES Secretariat maintained that the role of non-governmental organisations should be confined to capacity-building functions such as funding support, training, technical assistance and the provision of intelligence. Citing legitimate concerns that NGOs might inadvertently compromise customs or police operations due to insufficient coordination with relevant authorities, the Secretariat advised that such organisations should refrain from undertaking roles better suited to official law enforcement bodies. It regarded the TRAFFIC Network as the only NGO appropriately positioned to engage in such activities.

By the 13th Conference of Parties, it became evident that the T.I.G.E.R.S. database was failing. CITES argued that it was due to lack of resources that it was unable to enter all the information it received into T.I.G.E.R.S. database. At the meeting, the CITES Secretariat emphasised the submitting of at least two major items—a 'detailed information on significant cases of illegal tra (CITES, 2007b). In response to these shortcomings, NGO experts from the Asian Conservation Alliance, together with a research team from the United Nations University, initiated the development of the Wildlife Enforcement Monitoring System (WEMS)—a comprehensive geo-spatial decision-support platform designed to incorporate information from civil society sources (see Sect. 2.5).

An end to the T.I.G.E.R.S. database came during the CITES COP14, where the Secretariat admitted that T.I.G.E.R.S. has failed to meet its purpose. CITES declared that the submission of information to the Secretariat, suitable for input to T.I.G.E.R.S., has always remained haphazard and incomplete, and the database could never reach its potential (CITES, 2007b). According to the Secretariat, the failure was not just that the Parties were not submitting enough information but also because of the shortage of resources to input the large backlog of data that it received into the database. CITES also admitted that all the databases, including that of INTERPOL and World Customs Organisation, faced similar challenges (CITES, 2007b). The ETIS database operated for CITES by TRAFFIC was the most comprehensive one only because of the considerable effort that had been made to collect data proactively and encourage their submission (CITES, 2007b). The Secretariat expressed its disappointment that the Parties had not chosen to make more use of the ECO-MESSAGE system and, in turn, ICPO-INTERPOL's database or T.I.G.E.R.S. Had they done so, a much better record on the perspective would be available about illicit trade, wildlife crime and the persons involved (CITES, 2007a).

To overcome the challenges in enforcement information-sharing, the CITES Secretariat convened a meeting at the US Fish and Wildlife Service Forensics

Laboratory in Ashland (Oregon) from eighth to tenth June 2009. The meeting, which involved high-level CITES law enforcement experts (mostly practitioners), held several closed and open door sessions where they brainstormed on ways to improve the gathering of data on illicit wildlife trade. Based on the discussions, the experts recommended the development of a global database modelled upon EU-TWIX (developed by TRAFFIC and used by the enforcement community in the European Union; also see Chap. 5 Section 5.3.2). The group further noted that though the WEMS developed by UNU had good potential, it had to be tested in the field before being endorsed by CITES. This recommendation was later ratified at the 58th Standing Committee meeting (CITES, 2009) and at the CITES COP 15 meeting (CITES, 2010a). The CITES COP 15 meeting also recommended the establishment of a 'working group' for looking into the following:

(a) The design and the implementation structure of the database.
(b) Seek external funding to enable the group to conduct its activities, assisted by a relevant consultant if appropriate.
(c) Report to the Standing Committee at its 61st and 62nd meetings on the progress of the working group.

A brief meeting of interested Parties and organisations was then convened by the Secretariat during COP15, and a working group was subsequently established with the following members: Canada, Israel, New Zealand, United States of America, the European Union, INTERPOL, the World Customs Organisation, UNEP-WCMC, Environmental Investigation Agency (EIA) and TRAFFIC. The group was chaired by a Customs official from France. The group conducted its discussions by email and focused on issues such as: the structure of the database, the type of information it would contain, how information should be submitted and how such a database should be maintained (CITES, 2011a, 2011b). But during the sixty-first meeting of the Standing Committee, neither the Parties nor the Secretariat had the confidence to maintain a database on illegal trade, regardless of the nature of such a database (CITES, 2011c). The Secretariat admitted that it had no capacity to take on such work and other organisations represented in the working group were not ready to do it for free (CITES, 2011c). With this note, the Secretariat asked the Working Group on the 'Gathering and Analysis of Data on Illegal Trade' to postpone its further deliberations. A final decision on the database came during the 62nd Standing Committee meeting held in July 2012, in Geneva where the CITES Secretariat confirmed that it did not have the means to support a separate database on illegal trade. It requested CITES Parties to make better use of existing intergovernmental databases such as those maintained by INTERPOL and WCO (CITES, 2012b).

Another development that took place during CITES COP 15 (November 2010) was the formation of the International Consortium on Combating Wildlife Crime (ICCWC)—a joint partnership between (CITES), ICPO-INTERPOL, United Nations Office on Drugs and Crime (UNODC), World Bank and World Customs Organisation (WCO). The purpose of this consortium was to increase collaboration and cooperation among the enforcement agencies (CITES, 2010b, 2013c). The deliberations at CITES and part of this research are well enumerated in Chandran et al. (2021).[11]

Following up on the earlier decisions, at the 16th Conference of Parties held in Bangkok, CITES Parties further agreed to reduce the burden of reporting on illegal trade by adding some columns to the annual report rather than submitting a special report (CITES, 2013d).

As an overview, it is clear that the complexities of enforcement information collection and sharing existed since the formation of the convention. A direct approach to address the problem would lead to shaming one country while lauding the other, thereby creating objections among the Parties and causing reluctance to provide information to a central system based at the CITES Secretariat. This led to the failure of the illegal wildlife crime database (T.I.G.E.R.S) managed by the CITES Secretariat. On the contrary, unlike T.I.G.E.R.S, ETIS and MIKE databases originated as a precondition to trade in ivory. The reports and analysis from these databases are mandatory for countries to agree upon a decision on trade. Hence, the Parties provided generous contributions to keep the system functioning though it has often been criticised as not providing adequate analysis(Christy, 2012). The final CITES position on enforcement information-sharing essentially returned to an earlier decision made in March 1994, when the Secretariat reported mixed reactions from Parties regarding the proposal to establish a new, dedicated enforcement mechanism within CITES. At that time, it was acknowledged that leveraging existing intergovernmental enforcement mechanisms presented a more effective path forward (CITES, 1994a). TRAFFIC remains the only non-governmental organisation that CITES has formally entrusted with the management of enforcement-related databases. This may be attributed to the long-standing institutional relationship between CITES and IUCN—the organisation instrumental in the creation of the Convention itself. This dynamic becomes clearer when examining CITES' position on grassroots civil society participation, particularly in relation to the WEMS initiative.

2.5 Genesis of Wildlife Enforcement Monitoring System (WEMS)

As it was becoming increasingly evident that the compliance system of CITES was failing and the process of information-sharing was getting more complex and inefficient, researchers at UNU envisaged the idea of a civil society-based information-gathering process through an arrangement with the Asian Conservation Alliance (ACA). Before having an agreement on WEMS, ACA had several consultations

[11] Remi Chandran, Shankar Prakash Alagesan & Walter T. de Vries (2021) CITES enforcement information-sharing—*if you don't know where you've come from ... you don't know where you're going*, Journal of International Wildlife Law & Policy, 24:3–4, 343–365, DOI: https://doi.org/10.1080/13880292.2022.2043410

with their members in China, Thailand and India. The idea agreed upon by both UNU and ACA were as follows:

- An ACA-member NGO with significant experience in CITES processes will act as a national focal point, collecting and compiling information from governments on the basis of an information disclosure law (Freedom of Information Act, FoIA), and evaluate and verify the information before inputting it into WEMS (see Fig. 2.6).
- By doing so, WEMS can act as a geospatial decision support system accessible to a closed group of stakeholders participating in the project and involved in addressing enforcement and compliance on CITES, including governments, UN agencies, scientists and civil society.
- An open map portal displaying the illegal trade route and species involved in the trade will be the channel for public awareness on wildlife crime.

Based on the above agreement, on October first 2005, UNU and ACA signed a memorandum of understanding to carry out the project as a joint initiative. The particular feature of WEMS-NGO model was that it enabled grass-root NGOs to enter and share information on wildlife crime. The role of UNU was to provide a database infrastructure to which NGOs could input data. UNU researchers took the responsibility of analysing the data. In the same month, the International Fund for Animal Welfare (IFAW) donated US$30,000 to ACA of which US$15,000 was provided to UNU for building the database. On 18th November 2005, a consultative meeting was held among the ACA partners to discuss the design process and data collection strategy. In order to ensure interoperability with other enforcement-related databases, the WEMS data format followed that of an ECO-MESSAGE (CITES, 1995).

The user requirements were defined in consultation with Asian NGOs who were involved in enforcement-related work. After completing the consultation, a decision was made to start with Japan as a pilot phase and then expand it to other regions. ESRI was contacted on 14th December 2005 to procure GIS software. ESRI provided grants worth US$ 50,000 for the WEMS project to kick-start the WEMS initiative. Based on the NGOs requirements, a prototype of WEMS was developed with a spatial component that allowed the visualisation of the ECO-MESSAGEs on a wildlife crime map. The software design was based on the traditional waterfall model and included several steps: requirement analysis, software design, implementation, verification and maintenance.

By February 2006, while the prototype neared its completion, the data collection was shaping up too. Japan Wildlife Conservation Society (JWCS), one of ACA's partner NGOs, systematically recorded and verified all the seizures that had taken place in Japan since 2002 by compiling information from Japan Customs and Japanese Police using the Freedom of Information Act. This data was then sent to UNU for testing the prototype. During the prototype phase, only data from Japan was used for testing. The data was then entered into a Microsoft Access database. The database was linked to a web-based ARC-GIS mapping facility (ARC-MAP) which displayed the trade routes, quantified information on the species traded and

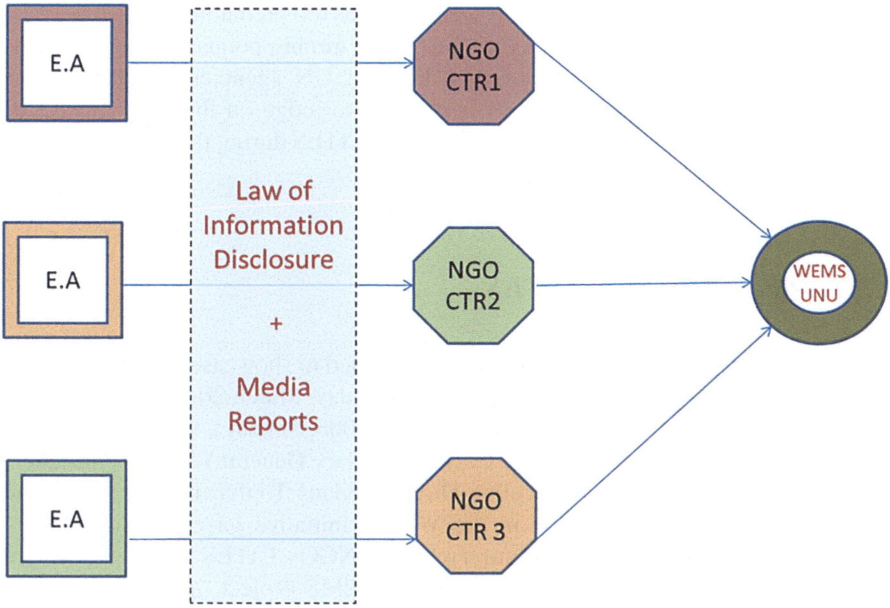

Fig. 2.7 Data sharing model of WEMS NGO prototype—(*EA* Enforcement Agency, *CTR* Country). The figure above describes the information collection process in the WEMS-NGO prototype. The information is first collected either through direct communication with the national enforcement agency (EA), by appealing to an existing Freedom of Information Act or by consulting media reports in each Country (CTR) through an NGO focal point (NGO) which enters the information to the WEMS system after a preliminary data verification regarding credibility of the data. The arrows show the transfer of information into the WEMS system placed at UNU (WEMS-UNU). The user requirements were prepared by UNU and ACA

the number of seizures. Though the information contained in the database was restricted to the WEMS stakeholders, the map was designed for the purpose of public awareness and was supposed to be open to the public (see Fig. 2.7).

The map in Fig. 2.5 shows the output from the WEMS prototype where it outlines the illegal pet trade into Japan. The data was collected by the Japan Wildlife Conservation Society (ACA partner NGOs) from the Japanese Customs and Police Agency. It also contained a querying unit, which the public could use to query on illegal trade numbers, type of species, etc.

The key point to note here is that the user requirements of WEMS invariably were the requirements of the Asian NGOs. Hence by default, the system design allowed the Asian NGOs to be the primary users of the system and the national enforcement agencies as the secondary users (enforcement officials can view the data only when the NGO has entered it). Another crucial aspect that should be taken note is, during the whole process, the WEMS team did not inform the CITES Secretariat or INTERPOL about the development of WEMS though it was directly related to the convention and the work of INTERPOL's Environmental Crime unit. According to UNU researchers, WEMS was at a prototype stage and due to the experimental nature of the database, it seemed it was too premature to inform external agencies. This can be considered as one of the turning points leading to confusion/conflict between the works of two different UN agencies in addressing a common problem (see sect. 2.5.1). It is difficult to judge on the consequence of what would have happened if UNU had informed CITES during the earlier stages of WEMS development.

2.5.1 Resistance from CITES

Once the prototype became functional, UNU decided to showcase the system to the public. A press conference was scheduled on Monday, March 20, 2006, at UNU in Tokyo. A day before the conference (March 19th 2006, Sunday), UNU Rector Hans Van Ginkel received a letter from the CITES Secretary General Willem Wijnstekers which was copied to UNEP Executive Director Klaus Töpfer, reasoning that the high level of NGO involvement in the WEMS initiative gave rise to concern. Referring to the fact that the information relied on NGOs, CITES Secretary General warned that grass-root NGOs involved in the WEMS project *'appear to support strongly protectionist wildlife conservation and animal welfare policies. Their stated aims are not only to promote law enforcement but also to reduce commercial wildlife trade overall. This latter aim is not consistent with the Convention, which regulates but does not prohibit wildlife trade…'*.

The main concern of CITES was: Asian Conservation Alliance (ACA) and IFAW represent preservationist positions and their objectives stood different from the stated objectives of those of the Convention. The letter requested an immediate postponement of the press conference on WEMS. Obliging to the synergy between UN agencies, UNU postponed the press conference until a clarification was made between all involved stakeholders, including the CITES Secretariat.

2.5.2 Redesign of WEMS

Following the objection, CITES and UNU met during a series of consultative meetings to discuss about the WEMS initiative. During all these meetings, the WEMS project went through an extensive consultation and scrutiny process (Table 2.1). The final outcome of the meetings was to exclude NGOs in enforcement information-collection, as transboundary information-sharing by NGOs could question the sovereignty of states (Chandran et al., 2011).

Following the discussions, two major decisions were made by UNU at that time (after June 2007):

- Consult with governments who are willing to take up WEMS project as a national government-led initiative.

Table 2.1 Details of the consultative and scrutiny meetings of WEMS system

Meeting	Location and date	Organiser	Purpose
CITES Mekong regional meeting	Kunming, China 4th–sixth July 2006	CITES secretariat	Showcase WEMS to Asian regional partners especially in using WEMS as a tool in transboundary information-sharing in the Mekong region. This meeting led to subsequent meetings with ASEAN-WEN secretariat for considering WEMS implementation in the ASEAN region.
INTERPOL wildlife working group (IWWG) meeting,	Beijing, China—25th–28th July 2006	INTERPOL	Evaluation of WEMS initiative by CITES and INTERPOL enforcement officials.
UNU-CITES secretariat meeting	Tokyo—third August 2006	United Nations University	Showcase WEMS to the diplomatic corps and selected public officials. The meeting was attended by the CITES enforcement officer, officials from US embassy, Ministry of Environment of Japan and United Nations University.
CITES—COP 14 meeting Hague	Hague, June 3–15, 2007	CITES secretariat	Showcase WEMS to CITES parties. Government of India participated in the side event along with UNU, highlighting the importance of implementing WEMS.

– Contact United Nations Environmental Program (UNEP) to support the implementation work.

During a conference call on August seventh, 2007 with UNEP, it was then decided to start with one country in the pilot phase. This then required restructuring the database from a transnational database to a national database. Subsequently, UNU chose India as a possible candidate country for testing WEMS as a national geospatial decision support system to assist national enforcement agencies involved in addressing wildlife crime. This involved bringing together multiple central ministries responsible for administering police, customs and forest in India. Prior to the negotiation process with the Government of India, UNU made two changes in the organisational structure of the WEMS project;

- Firstly, UNU discontinued its partnership with ACA.
- Secondly, on October 2007, UNU moved the project to the e-governance unit of UNU—UNU-Institute for Software Technology (UNU-IIST) based in Macao.

Although the Government of India welcomed the initiative, WEMS implementation at a national level in India turned out to be more challenging than expected. The intention of the Government of India was to first host WEMS at its wildlife research institution, Wildlife Institute of India (WII), which had the expertise and capacity to operate and train officials on the WEMS project. However, by June 2007, the Government of India had also established the Wildlife Crime Control Bureau (WCCB)—a dedicated organisation focusing exclusively on wildlife crime related

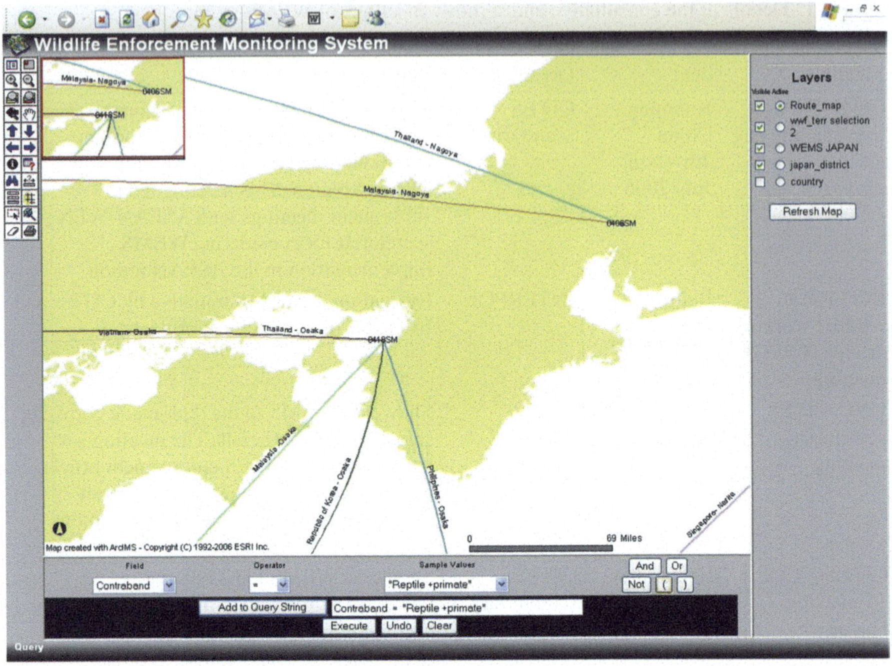

Fig. 2.8 Open map portal from WEMS-Asia Prototype

matters and headed by an officer from the Indian Police Service (IPS). Though both WII and WCCB were under the regulation of the Ministry of Environment and Forests (MoEF) of India, WCCB was not in favour of a research institute being a repository of wildlife crime information. At the same time, WCCB did not have the manpower or infrastructural capacity to host the system. This led to prolonged negotiations on who will be the custodian of WEMS system–WCCB or WII. The confusions surrounding the ownership led to a stalemate in the decision-making process. Meanwhile, disagreements among the WEMS project managers at UNU on how to operationalise WEMS in India as well as bureaucratic politics and turf fights during the design of the pilot testing eventually led to insuperable challenges in terms of coordination with WCCB and MoEF-India. The final outcome was a stalemate on a decision regarding the project.

By 2010, UNU had almost given up its hope of moving ahead with the project. The UNU project team presented WEMS at CITES COP15 where it was conceptualised as a working model (Fig. 2.8) for regional bodies like the Association of Southeast Asian Nations Wildlife Enforcement Network (ASEAN-WEN) and LATF as well as for national governments. However, WEMS was not endorsed by the CITES Secretariat as it had already made a decision to implement EU-Twix globally (Fig. 2.9).

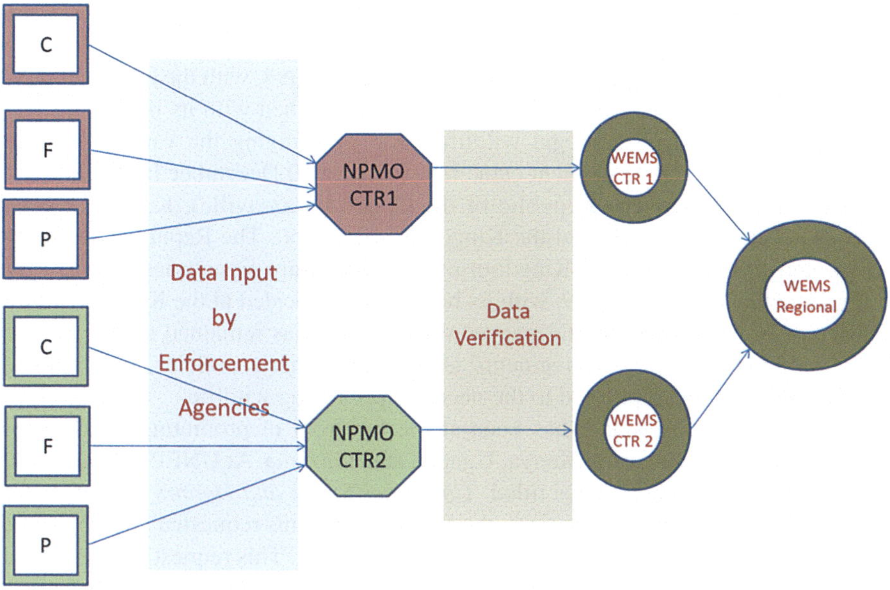

Fig. 2.9 Suggested working model for WEMS regional (ASEAN and LATF). The above figure describes the information collection process in the revised WEMS prototype. Here the WEMS focal point (national focal point responsible for administering the WEMS system at a national level) is based at the Wildlife division or CITES Management Authority of each country (CTR) that acts as a National Project Management Office (NPMO). The WEMS focal point then coordinates the information input from the customs (C), Forests (F) and Police (P) who are provided direct access. The arrows show the flow of information to the WEMS regional database

The project almost ended in 2010 at UNU-IIST. However, a swift change happened soon after CITES COP 15 when the Lusaka Agreement Task Force (LATF) contacted UNU to discuss implementing WEMS in Sub-Saharan Africa.

2.5.2.1 Lusaka Agreement Task Force (LATF): An Overview

The formation of Lusaka Agreement Task Force (LATF) in 1992 came at a time when the international ban on ivory trade was in place. One criterion for lifting the ban was a requirement for good domestic law enforcement by both the consuming and trading nations. African countries that were not prioritising wildlife trade also required support from transboundary countries on law enforcement in protecting their wildlife. To meet this requirement, in 1992, Sub-Saharan African countries formed coalitions and recognised wildlife law enforcement as a transboundary priority. The idea for the formation of LATF emerged during a conference hosted by the Zambian Ministry of Tourism at Lusaka, where senior wildlife law enforcement officers from eight African countries met to propose an African Task Force in order to coordinate cross-border investigations in close liaison with national law enforcement agencies (UNEP, 2005). At that meeting, they drafted an agreement aimed at providing a legal framework for the proposed Task Force, and named it the Lusaka Agreement on Co-operative Enforcement Operations Directed at Illegal Trade in Wild Fauna and Flora (UNEP, 2005). On 8 September 1994, with the UN Secretary-General acting as its depositary, the LATF was established with its main objective to collectively act against illegal wildlife trade by bringing the various wildlife agencies together. The Agreement came into force on 10 December 1996, and constituted seven Parties: The Republic of the Congo (Brazzaville); Kenya; Liberia; Tanzania; Uganda; Zambia and the Kingdom of Lesotho. The Republics of South Africa and Ethiopia, and the Kingdom of Swaziland, are signatories. It was officially launched on 1 June 1999, with its headquarters located at the Kenya Wildlife Service in Nairobi. Since its formation, the coalition has remained stable and the Parties, though having disagreements about each other's policies, have stayed together and remain committed to the needs of protecting wildlife.

The Zambian government has been at the forefront of promoting LATF, supported by the governments of Kenya, Uganda and Tanzania. At UNEP's Conference of the Rhinoceros Range States titled '*Consumer States and Donors on Financing the Conservation of the Rhinoceros*', the four governments requested UNEP to take up a coordinating role in negotiations for the Agreement. This request was endorsed unanimously by other countries present. UNEP responded by setting up a 'Coordinating Secretariat' to facilitate the negotiating process and by committing US$200,000 to get the project started (UNEP, 2005).

Role of NGOs in the Formation of LATF

Although the LATF was formed through a multilateral arrangement involving African states, some scholars (Reeve, 2014) argue that LATF was triggered by NGO action. According to Reeve (2014), Nick Carter—a Namibian NGO advocate until his death in 2000—played a crucial role in lobbying for the formation of LATF which later won him the Goldman Sachs award (Reeve, 2014). This argument can be further substantiated by the fact that for a period of time, LATF was a project of the David Shepherd Wildlife Foundation (DSWF, 2014). LATF also received support from conservation NGOs such as: Friends of Animals; The Humane Society of the United States; The International Fund for Animal Welfare (IFAW) and The Environmental Investigation Agency (EIA) (UNEP, 2005). This coalition with NGOs has weakened drastically in recent times as NGOs shifted to funding governments directly rather than supporting a network.

2.5.3 Adoption of WEMS by LATF in Sub-Saharan Africa

After hearing about WEMS during the CITES enforcement expert group meeting[12] and at the 16th CITES COP meeting (CITES, 2010a, 2010b) in 2010, LATF contacted UNU to discuss the procedure of implementing WEMS in Africa as a pilot

[12] The CITES Enforcement Expert Group meeting held in Ashland, United States, discussed the scaling up of WEMS as a global database; however, they stopped short of endorsing it as a global model (CITES, 2009).

project. By then, the administration of WEMS was moved to UNU-Institute for Advanced Studies (based in Japan). At the time of request, WEMS was still in its prototype stage. For scaling it up into a regional transboundary geospatial database, WEMS required certain security protocols which would allow countries to securely share information among them (Chandran et al., 2011).

Learning from earlier failures, this time UNU partnered with several research institutions to strengthen its implementation mechanism in Africa. UNU, along with Harvard University's Center for Geographical Analysis and ITC-University of Twente, formed a working group to look into the research as well as the implementation process of WEMS. Unlike in Asia, the African implementation mechanism was backed by a strong research community from Europe, Asia and North America.

The funding for the pilot implementation was provided by IFAW directly to LATF. UNU used its own resources to develop the ICT cloud infrastructure (based at UNU-campus computing centre) where WEMS was later hosted. LATF also sourced funding from UNEP to convene workshop and training programmes for initiating WEMS in Africa.

Following several workshops and meetings to gather the user requirement from the enforcement agencies (primarily, Customs, Forests and Police) in Kenya, Uganda, Tanzania and Congo, WEMS finally got implemented—nearly 6 years after it was first conceptualised at UNU in Tokyo (Fig. 2.10).

On July 2011, the Minister of Forests and Wildlife of Kenya Dr. Noah Wekesa inaugurated the launching of WEMS in four Sub-Saharan African countries—Kenya, Tanzania, Uganda and Congo Brazzaville. Speaking at the occasion, Dr. Noah Wekesa said WEMS will 'mark a [new] beginning in the history of Africa,

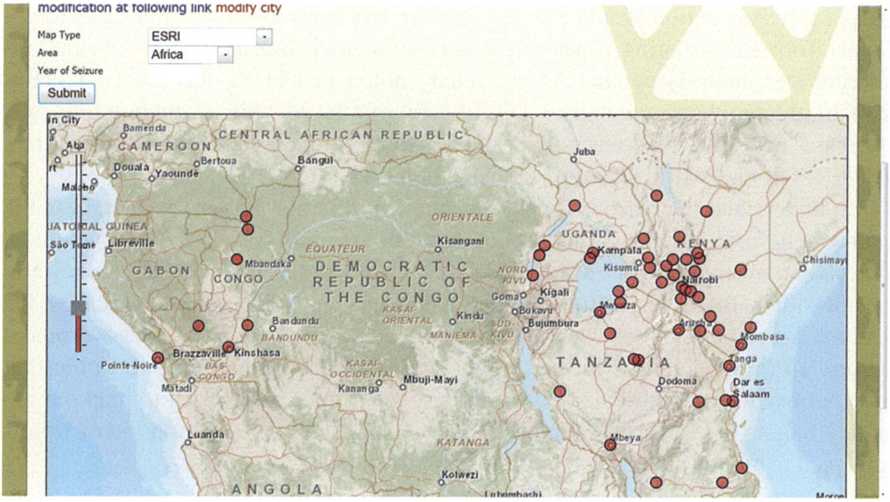

Fig. 2.10 WEMS Map output from the regional model. The map above shows the output of WEMS-Africa: the locations of wildlife seizures. The information was input directly by the respective enforcement official and coordinated by the WEMS focal point in each country

where we show the world an example of good governance' (UNU, 2011). President of Kenya, Mwai Kibaki, also expressed his support for such a system (Kenya, 2011).

Since then, WEMS transformed from a prototype to a functional transboundary geospatial decision support system used by the four Lusaka Agreement countries. The countries in the pilot phase met for two training sessions and the database was restructured in such a way that WEMS will act as a national and regional database (Chandran et al., 2011). While all the cases are input into WEMS by the participating country, only certain information which has transboundary relevance is shared with LATF. All the countries agreed to this clause and WEMS became fully functional in 2012. As of 2014, the pilot countries who participated in WEMS had shared around 540 cases across borders (LATF, 2014, 2015). A report by UNU considers the WEMS implementation as a success case in Kenya, Uganda, Tanzania and Congo Brazzaville (Chandran et al., 2013).

Though WEMS was implemented in four African countries, its approval by the CITES Secretariat remained critical. Until 2013, the CITES Secretariat hesitated to endorse the project, claiming that it would create confusion in the ongoing work of the International Consortium on Combating Wildlife Crime (ICCWC). According to CITES Secretariat (mentioned during CITES COP16 side event in March 2013):*'WEMS should not be promoted or used as an information system which might somehow overlap with, duplicate, or divert attention and resources from, efforts to improve the submission of wildlife law enforcement information through the INTERPOL and WCO information systems'*.

This position of CITES changed after the Tokyo Conference on Combating Wildlife Crime held on March 3, 2014, in Tokyo, where a broader and conditional acceptance of WEMS was agreed upon (see Chap. 6).

Although CITES has given several reasons for objecting to WEMS, it is not clear why a UN convention would prevent another UN body from facilitating enforcement information-sharing, especially when challenges in transboundary information-sharing and analysis persist. As previously noted in CITES documents (CITES, 2007b), the problems which the CITES Secretariat faced while maintaining its own database are still experienced by the existing law enforcement agencies like INTERPOL and World Customs Organisation (WCO).

To understand the reasons for this scepticism, it is first important to analyse the policy beliefs within the wildlife policy subsystem within Asia and Africa where on one end an objection to WEMS halted its NGO-based data collection in Asia, while on the other, there was a full-fledged acceptance in Sub-Saharan Africa by the Lusaka Agreement Parties when the data collectors were replaced with government actors. To investigate this further, a two-step procedure is carried out. In the first step, an extended empirical study is carried out in Asia and Africa to map out the policy beliefs (see Chap. 4). And, as a subsequent step, the coalition interaction is analysed using the advocacy coalition framework (see Chap. 5).

2.6 Conclusion

The trajectory of the Convention on International Trade in Endangered Species of Wild Fauna and Flora (CITES) has been one of significant transformation, characterised by both achievements and setbacks. Since its inception in the early 1970s, CITES has endeavoured to navigate the complex intersection of environmental protection, trade regulation and political will. Over time, the Convention has faced an ever-expanding web of challenges, with global trade, shifting political interests and the growing need for sustainable practices continually shaping its evolution.

In its first two decades, CITES focused on operationalising the Convention and fostering collaboration, particularly with international non-governmental organisations (NGOs) such as IUCN, WWF and TRAFFIC. These partnerships provided vital expertise and resources, proving essential as the scope of the Convention widened. Yet, despite early optimism, financial limitations and institutional capacity proved formidable obstacles. The decision to enlist the expertise of IUCN was pivotal, ensuring the Convention's sustainability and reinforcing its role in global conservation.

By the late 1970s, the Convention began to experience an ideological rift, driven by the entry of countries with differing priorities, particularly in East Asia. The debate surrounding species listings and the concept of 'sustainable use' shifted the Convention's approach from preservationist values towards a more utilitarian stance. This ideological tension between conservation and commercial interests has remained a defining feature of CITES ever since.

As CITES expanded, so too did its logistical and political challenges, especially regarding compliance and transparency. Over time, the optimism of the early CoPs gave way to growing reluctance to report infringements, as political pressures and national interests began to obscure the Convention's collaborative mandate. The inability to effectively monitor and report wildlife trade led to frustration among international observers and the Secretariat, as many Parties, particularly in the developing world, struggled with enforcement. By the 1990s, CITES faced its most significant test with the African elephant crisis, revealing deep inefficiencies in the enforcement of its decisions.

In response, the establishment of the Wildlife Enforcement Monitoring System (WEMS) in 2005 marked a pivotal moment in addressing these enforcement challenges. Developed by the United Nations University (UNU) in collaboration with the Asian Conservation Alliance (ACA), WEMS sought to improve data collection on illegal wildlife trade through grass-roots NGOs. This system empowered these organisations to become national focal points, allowing for the real-time tracking of wildlife crime and fostering greater transparency and accountability among key stakeholders, including governments and UN agencies. However, resistance emerged from within CITES itself, due to concerns about the diverging objectives of NGOs, which were often more preservationist in their stance. This ideological clash highlighted the challenges of reconciling conservation efforts with the practical realities of international trade regulation.

Despite these tensions, the WEMS project represents a significant milestone in the effort to enhance data sharing, transparency and enforcement. As CITES continues to evolve, it must build upon these lessons. The future of the Convention lies in its ability to adapt to the changing dynamics of global wildlife trade while maintaining a strong commitment to conservation. To achieve this, CITES should strengthen its partnerships with international agencies such as INTERPOL and the UNODC, leveraging initiatives like the International Consortium on Combating Wildlife Crime (ICCWC) for enhanced enforcement. The use of emerging technologies, including artificial intelligence and satellite tracking, will also be vital in improving monitoring and enforcement systems.

Looking ahead, integrating the CITES Trade Database, Wildlife TradeView and WEMS into a cohesive Digital Public Infrastructure model would represent a significant step forward. This model would facilitate the seamless exchange of data between governments, NGOs, scientists and the public, improving transparency, accountability and real-time monitoring of wildlife trade across borders. Such an integrated platform would empower all stakeholders to collaborate more effectively, enhancing enforcement efforts and fostering greater global cooperation in the fight against illegal wildlife trade. In this framework, digital tools would be central to enabling better decision-making, more robust reporting mechanisms and greater public awareness.

CITES must also continue its emphasis on capacity building, particularly in developing countries, to ensure that all Parties have the necessary tools and knowledge for effective enforcement.

Looking ahead, the success of CITES will depend on its ability to bridge the divides between political interests and conservation goals, ensuring that enforcement mechanisms are more effective and inclusive. Its legacy will not be defined by the number of species saved or the volume of trade regulated, but by its ability to adapt to the evolving dynamics of global conservation. As CITES moves forward, it must remain focused on the central mission of protecting biodiversity, while recognising the complexities inherent in international environmental governance. Only through innovation, collaboration and a commitment to inclusivity can CITES evolve into the effective and dynamic convention our planet urgently needs.

Chapter 3
The Workings of International Wildlife Politics and Policy Domains

3.1 Introduction

As we have seen in the previous chapter, conflicting ideologies on several policy issues between national governments, NGOs and academic institutions have made CITES decision-making a complex process. Solving such complexity by means of a rationalistic policy approach alone also seems inadequate as policy processes are influenced by macro- and micro-level changes within and external to the policy domain or subsystem. In order to understand the policy interactions within the Wildlife–Trade policy subsystem,[1] this chapter records the policy process within the subsystem spanning more than a decade by using a theoretical framework of policy change based on the beliefs and behaviours of individual actors or groups of actors working within the policy domain. This is specifically important as no study has conceptualised the beliefs and policy process within CITES over time (during a time span of more than a decade) as well as the role of coalitions and other external factors in influencing the Wildlife–Trade policy subsystem.

The Advocacy Coalition Framework (ACF) (Sabatier, 1988) clarifies policy processes at the international level and pays attention to policy coalitions. The application of ACF to MEAs is not new. Sewell (2005), for example, used ACF to highlight coalition behaviour within the UNFCCC climate policy context and examined the climate policy processes at the international, national and subnational levels, including the nature of cooperation and coordination required both within and between these levels. A similar study was carried out by Ganguly (2010) who applied ACF to the case of the Convention on Biodiversity (CBD). However, more recent work by Ingold and Varone (2011) has pointed to factors that explain policy change within

[1] When two policy domains—such as Wildlife policy and Trade policy—intersect, the specialised unit concerned solely with matters pertaining to both is referred to as the Wildlife–Trade policy subsystem.

R. Chandran, *The Wildlife Enforcement Monitoring System*, Professional Practice in Governance and Public Organizations,
https://doi.org/10.1007/978-3-031-97084-9_3

MEAs, such as institutionalised veto points and the strategic behaviour of policy brokers. Although, theoretically, CITES processes can be illuminated through ACF, it alone may not be sufficient to show the external dynamics influencing a policy subsystem, especially within a multilateral context, where decisions are not just influenced by the national policies of a state, but also and perhaps more so where certain (weak) states are influenced by lobbying groups, international NGOs and the vested interests of other (powerful) states. In order to fill this void, this book refers to the work of Keck and Sikkink (1999), which explains how actors who fail at the national level seek an international platform to address their concerns.

The first part of this chapter (Sect. 3.2) describes the Advocacy Coalition Framework. The following section (Sect. 3.3) then describes its theoretical application to the CITES Convention. The final section (3.4) illustrates through a case study the interactions within and between the national and international policy subsystems, and how this transnational effect has influenced policy decisions over a decade. The chapter concludes by demonstrating the relevance of this theory to the context of CITES processes and its application in understanding the processes that took place in the development and implementation of the Wildlife Enforcement Monitoring System (WEMS).

3.2 Advocacy Coalition Framework: General Framework

The Advocacy Coalition Framework (ACF) helps to map out policy change over a longer period of time where actors within a particular policy subsystem can be aggregated into a number of advocacy coalitions, each including people from various state and non-state organisations that share a set of normative and causal beliefs while engaging in a coordinated activity.

There are three basic premises to ACF:

1. The process of policy change and policy-oriented learning requires a time perspective which is usually a decade or more. The focus on a decade as a time span comes directly from findings concerning the importance of the 'enlightenment function' of policy research (Sabatier, 1988). The importance of analysing policy for a longer period of time in order to obtain a reasonably accurate portrait of the success or failure of a programme is also stressed by Mazmanian and Sabatier (1983).
2. The second premise is that in a policy subsystem, actors from a variety of public and private organisations who are actively concerned with a policy problem (for example, wildlife trade) interact. This premise signals a departure from traditional notions of 'iron triangles' where policy concerns were limited to administrative agencies, legislative committees and interest groups at a single level of government. A policy subsystem includes actors at various levels of government who are active in policy formulation and implementation, as well as journalists,

researchers and policy analysts who play important roles in the generation, dissemination and evaluation of policy ideas.
3. The third premise of ACF is the conceptualisation of public policies through belief systems. Belief systems are based on a set of value priorities and causal assumptions of actors or of coalitions. Assuming that people get involved in politics at least in part to translate their beliefs into public policy, the ability to map beliefs and policies on to the same 'canvas' provides a vehicle for assessing the influence of various actors on public policy over time. ACF distinguishes three levels of belief systems, namely: (a) deep core; (b) policy core and (c) secondary aspects.

   - Deep core belief includes such things as: Basic ontological and normative assumptions about human nature; the priority of fundamental values such as liberty or equity and political support to a left- or right-wing party. It refers to the individual's underlying personal philosophy. This belief usually applies to all policy fields and provides the foundation for other more specific beliefs.
   - Policy core elements: These beliefs relate to a specific policy field or subsystem. It includes the basic strategies and policy positions required to satisfy deep core beliefs within the policy area of a subsystem and involves topics such as: whose welfare counts; the relative authority of governments versus the market; and the role of various actors in the policy subsystem.
   - Secondary aspects: These relate to particular aspects of the policy area and are narrow in scope. A secondary aspect refers, for instance, to the multitude of instrumental decisions and information searches that are necessary in order to implement the policy within a specific policy area.

According to ACF, deep core beliefs are normative beliefs, such as norms of fairness, that operate across most policy domains. Deep core beliefs are difficult to change but have the least salience concerning a particular policy issue (Henry et al., 2010). Policy core beliefs are normative commitments and causal perceptions related to an entire policy subsystem, such as the causes of a problem or the appropriate balance between economic and environmental goals (Henry et al., 2010). Policy core beliefs are fairly stable over time but are more relevant to a specific policy system. Secondary beliefs are related to specific geographic locations, policy proposals or information systems, and change more frequently in response to new information (Henry et al., 2010).

As the focus of ACF is on the policy subsystem, it is important to consider the various exogenous variables affecting the actors and resources of the subsystem. Sabatier (1988) classifies these exogenous variables as stable (Relatively Stable Parameters) and dynamic (Dynamic or External (system) events). Sabatier (1988) stresses that the two variables affect the coalition formation and the resources of the policy subsystem.

In the next subsection, the above-mentioned exogenous variables are explained in detail.

3.2.1 External Factors Affecting Policy Change within Subsystems

According to ACF, the policy subsystem is embedded in a wider system of events that provide coalitions within the policy domain with different constraints and opportunities. Sabatier (1988) calls this the 'external factors (or exogenous factors) affecting policy change within a policy subsystem' and classifies it as:

1. Relatively stable parameters—producing little or slow change.
2. External (system) events (also referred to as dynamic events)—producing rapid and unpredictable change.

Relatively Stable Parameters

During a policy process, certain factors that are independent of the discretion of the actors can influence policymaking. Such factors are related to:

(i) Basic attributes of the problem area: In most of the policy process, the attributes of the problem are almost fixed and there is little room to manoeuvre. For example, on decisions related to the management of forests and wildlife (or any Common Pool Resources—CPRs—in general), an outright decision based on a problem cannot be made. It requires discussion with the owners (governments/private) and action within the laws which define the management of the resources. This means that a solution can only emerge with the consent of the owner in charge of the resource. Although external factors (see next section on external subsystem events) influence the regulation, the rules concerning the management of resources are almost stable.

(ii) Basic distribution of natural resources: According to Sabatier (1988), the present (and/or past) distribution of natural resources strongly affects a society's overall wealth and the viability of different economic sectors, as well as many aspects of its culture and the feasibility of options in many policy areas. Here, policy shift takes place based on the availability of the resources. The position of African countries on ivory trade is a classic example of how decisions can be influenced by resources. Kenya, for instance, due to its lower population of elephants has resorted to a total ban in ivory trade and trophy hunting of elephants; whereas South Africa, with its rich elephant population, permits trophy hunting of elephants, and it has a pro-ivory trade position.

(iii) Fundamental cultural values and social structure: Cultural values and certain basic rights associated with a particular issue can make it difficult to solve the problem. Changing policies related to these issues may take several decades. Douglas and Wildavsky (1983), Thompson et al. (1990) and Hoppe (2007) support the influence of beliefs in politics while describing through the lens of cultural theory how culture influences normative or core beliefs. A more detailed explanation of cultural theory, and its application in identifying normative beliefs, is provided in Chap. 4.

(iv) Basic constitutional structure (or legal structure): In most political and legal systems, the basic legal norms are usually resistant to change once a law is approved.

External [System] Events

According to Sabatier (1988), policy subsystems are susceptible to frequent fluctuations over the course of years due to dynamic events which then serve as a major stimulus for policy change. For example, the 2007 financial crisis was one factor which influenced a large number of policy subsystems. Several bills that were passed had to be changed to accommodate the crisis. This also presents a continuous challenge to subsystem actors, especially for learning how to anticipate and then respond in a manner consistent with their basic beliefs and interests. These external events are mainly influenced by the following sub-factors:

1. Changes in socio-economic conditions and technology: Where unexpected socio-economic scenarios may reverse a previously well-adopted policy decision.
2. Changes in systemic governing coalitions: An individual resignation or change in mind-set could bring in changes to the coalition.
3. Policy decisions and impacts from other subsystems: The decisions and impacts from other policy sectors are amongst the most important of the dynamic elements affecting specific subsystems. For example, the Tohoku earthquake in Japan and the Fukushima nuclear plant disaster influenced other policy subsystems, including agriculture, fisheries and transportation.
4. Changes in public opinion: Public opinion on any key issue can influence a policy decision. However, this applies more to the case of countries where there is a democratic participation in the electoral process.

Sabatier recognises the importance of understanding two intervening variables influencing the policy subsystem, which is especially relevant within the context of cross-national policy research. These two intervening variables are:

Long-Term Coalition Opportunity Structure

Both the relative stable parameters and the dynamic events affect the constraints and opportunities of subsystem actors and the way they form long-term coalitions. The three factors that define long-term coalition opportunities are:

1. Degree of consensus needed for major policy change; For example, at CITES COP meeting's, procedural matters relating to the conduct of the business of the meeting are decided either by consensus or by a simple majority of the Party representatives present and voting. In case of proposals related to trading of a species, decisions are taken by a two-thirds majority of representatives present and voting.
2. Openness of political system: The degree of openness depends on: a) number of decision-making venues and b) the accessibility to those venues. Within CITES Convention, there are several venues where decisions are made, such as the Standing Committee, Animals and Plants Committee, etc. These multiple access points allow Party participation for influencing decisions.

3. Overlapping societal cleavages: Here, cleavage is the division of voters into voting blocs based on personal or group priorities. Within CITES, voting is carried out in blocs by the European Union and by certain African states. Cleavages are also formed while voting on certain key species such as elephants and marine life.

The above three factors are very much dependent on stable and dynamic factors and can influence the way coalitions structure themselves.

Short-Term Constraints and Resources of Subsystem Actors
The exogenous factors also influence the resources of the subsystem actors. A shortage of resources can constrain the activities of actors in influencing policy change. For this reason, coalitions opt to share resources which will then make them partners in a coalition.

3.2.2 Policy Subsystem: Internal Structure

A policy subsystem is not just comprised of traditional policymakers but also includes journalists, analysts and researchers, who play important roles in the generation, dissemination and evaluation of policy ideas, as well as actors at other levels of government who are involved in policy formulation and implementation (Sabatier, 1988). The distinguishing features of a policy subsystem can be defined as follows:

- *Delimiting subsystem boundaries*
  According to Sabatier, a policy subsystem is usually well defined, and the coalitions have clear mandates. However, within the policy subsystem, there are latent actors who will be active only if they are convinced. In other words, a well-defined coalition constantly seeks support from latent actors.
- *Origins of new subsystems*
  A new subsystem can be formed when a group is dissatisfied or if it discovers important issues within a policy subsystem that are being neglected. This dissatisfied group will then form a new subsystem in order to draw attention to these issues.
- *Subsystem actors: advocacy coalitions and policy brokers*
  Whatever their origins, subsystems normally contain a large and diverse set of actors who form coalitions based on shared beliefs. In addition, there is a category of actors, the 'policy brokers', whose main concern is to keep the level of political conflict within acceptable limits and to assist in finding some 'reasonable' solution to the policy problem. Some literature defines policy brokers as influential actors who play a significant role in defining the policy processes and outputs (Ingold & Varone, 2011). The role of policy brokers will be described more in the latter part of the book.

Policy-Oriented Learning

ACF also stresses *policy-oriented learning* where coalitions or members of coalitions alter their thoughts or behavioural intentions as a result of experience or due to new information (Sabatier, 1988). Here, the members of various coalitions seek to better understand the world in order to further their policy objectives.

Assessing *policy-oriented learning* however requires deep involvement within the policy process as such shifts cannot be captured easily without being a long-term 'insider' within the policy subsystem. Hence, in this book, *policy-oriented learning* is outlined while describing the policymaking process on WEMS in Asia and Africa.

3.3 Advocacy Coalition Framework: Application within the Framework of the CITES Convention

In the introductory section of this chapter, the application of ACF in understanding the policy process of MEAs was outlined. Within the context of an MEA, the application of ACF to CITES also remains significant. However, studies related to the understanding of CITES processes by using ACF are limited. Although Steinberg (2003) and Arnold (2003) use ACF within the CITES context, their focus has been primarily on the national policy subsystem. On the other hand, CITES decisions are not just based on national agendas, but also include a multitude of factors involving: natural resource distribution at a global level; consensus between different actors at the national and international level; and the inherent cultural contexts of the actors that shape their beliefs and values on trade and conservation. For this same reason, studying the transnational policymaking process within CITES requires analysis of both the national and international wildlife policy subsystems. As a step towards this analysis, this section outlines the external (Subsection 3.3.1) and internal factors (Subsection 3.3.2) which influence the wildlife policy subsystem.

3.3.1 External Factors Affecting Policy Change to the Wildlife–Trade Policy Subsystem

Relatively Stable Parameters

According to ACF, relatively stable factors affect the beliefs and resources of the subsystem actors and limit the range of alternatives available to them. They also structure the problem, establish the rules and procedures for changing policy and reaching collective decisions, and broadly frame the values that inform policymaking (Sabatier, 1988; Weible & Sabatier, 2009). Within the CITES context, the following factors can be considered to be relatively stable:

(i) *Basic attributes of the problem area.*

The crisis in the trade and management of wild flora and fauna has been a long-term one, primarily due to extensive deforestation, species depletion and illegal trade over time. There appears to have been a continuous struggle, triggered by ideological difficulties on the nature of management—whether public, private or mixed management. The trade on wildlife has also re-emerged as an issue in the context of the debates on sustainable development and resource management. The exploitation of forests and wildlife had been rampant during the era of colonisation when there were fewer restrictions on the unwarranted use and trade of natural resources (see Chap. 2, Section 2.2). However, following decolonisation and with the emergence of the environmental movement in the 1960s, a new form of global governance emerged. Garrett Hardin's (Hardin, 1968) portrayal of the users of a common pool resource (CPR) later developed into the concept of a wiser use of resources through a hierarchical mechanism of control (Ostrom, 2009a). One important factor about CPRs such as wildlife and forests is that they yield benefits to both the primary (for example, forest dwellers and local communities) and the secondary dependants (people who use the resources for market or commercial purposes but who do not depend on them on a daily basis; for example, forest logging companies, professional hunters, etc.). This then leads to the problem where beneficiaries are hard to *exclude*[2] and consequently, each beneficiary's use of a resource system *subtracts*[3] units from that resource (Ostrom, 2009b, see Table 3.1).

In Table 3.1, Ostrom (2009b) uses two attributes—*exclusion* and *subtractability* (that range from low-high) in order to distinguish among four basic types of goods: toll goods (sometimes referred to as club goods); private goods; public goods and common pool resources (Ostrom, 2009b).

The struggle to classify exclusion of beneficiary and subtractability of use from a common pool resource is high (see Table 3.1) and is clearly visible in the Bern and Fort Lauderdale Criteria (see also Sect. 2.3 of Chap. 2). While Parties to CITES adopted the Bern Criteria in 1976, providing more impetus to preservation values of wildlife (a common pool resource), the Fort Lauderdale Criteria (which replaced the Bern Criteria), adopted in 1994, promoted a more utilitarian approach (Sands, 2003).

Table 3.1 Four basic types of goods (Ostrom, 2009a, 2009b)

	Subtractability of use		
Difficulty of Excluding Potential Beneficiaries		Low	High
	Low	Toll goods	Private goods
	High	Public goods	Common pool resource

[2] *Exclusion* relates to the difficulty in restricting those who benefit from the provision of a good or a service (Ostrom, 2009).

[3] *Subtractability* refers to the extent to which one individual's use subtracts from the availability of a good or service for consumption by others (Ostrom, 2009).

(ii) *Basic distribution of natural resources:*

A clear indication that the globally advocated control mechanism to manage nature was not working came from a UNEP study in 2007 (Siebenhüner, 2009) where it was estimated that about 23% of mammals and 12% of bird species were globally threatened and that populations of plants and animals had been declining since 1970. The Relative Global Index showed a decline of 14% for tropical forests, 35% for marine ecosystems and 50% for freshwater ecosystems. Most of these losses occurred in developing countries, where species diversity is the highest (Siebenhüner, 2009). This supports recent IUCN statistics (IUCN, 2014) which describe habitat loss and species extinction as amongst the biggest threats to large mammals, linking deforestation, urbanisation and trade as the three main factors leading to extinction of species (see Fig. 3.1).

The Red List Index (RLI) of IUCN (see Fig. 3.2) shows similar trends. It indicates a rapid decline in corals and amphibian populations followed by those of birds and mammals (IUCN, 2014). These statistics cast a shadow on the effectiveness and relevance of the CITES Convention, the only convention with the 'teeth' to sanction states for non-compliance.

(iii) *Fundamental cultural values and social structure:*

As mentioned in Chap. 2 (Sect. 2.4), conflicting perspectives on conservation have been in existence since the formation of multilateral environmental agreements. The three main beliefs involved in conservation debates are held by: the proponents of wildlife trade (Anthropocentrism); the opponents of wildlife trade (Ecocentrism,[4] Partridge, 1984); and actors who believe in a middle path balancing trade and conservation (Hierarchism, see Thompson et al., 1990). When the proponents of trade suggest that wildlife trade can bring additional income to

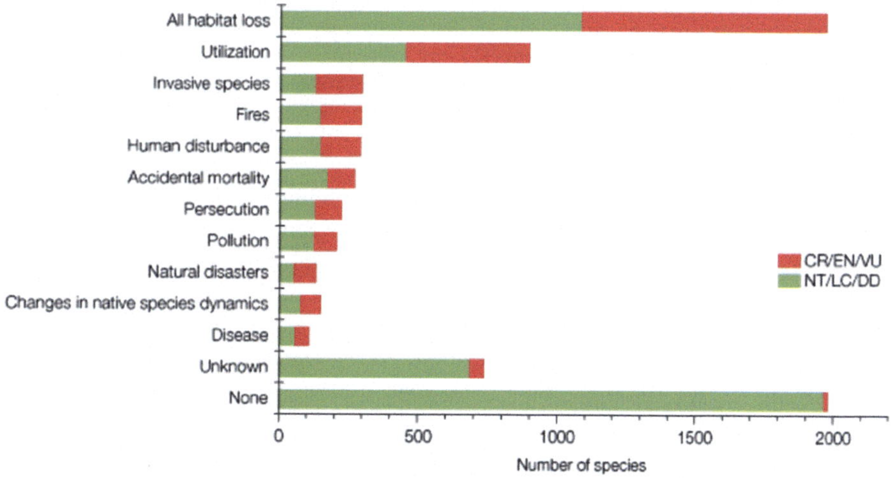

Fig. 3.1 Major threats to mammals (Source: IUCN (2014))

[4] Ecocentrism is also referred to as 'biocentrism' by some authors (see Partridge, 1984).

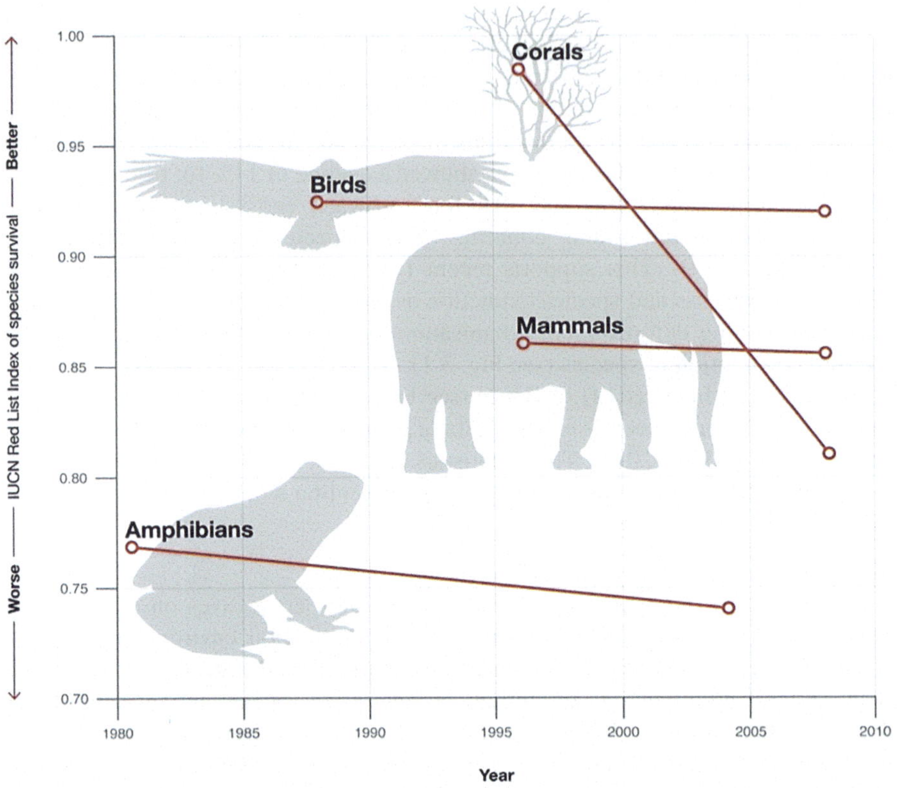

Fig. 3.2 Red List Index values (RLIs) for reef-forming corals, birds, mammals and amphibians (IUCN, 2014). The figure indicates that coral species are moving towards an increased risk of extinction most rapidly, while amphibians are, on average, the most threatened group. An RLI value of 1.0 equates to all species qualifying as Least Concern (i.e. not expected to become extinct in the near future). An RLI value of 0 equates to the species group becoming extinct. A consistent RLI value over time indicates that the overall extinction risk for the group is constant. If the rate of biodiversity loss was reducing, the RLI would show an upward trend (Source: IUCN, 2014)

resource-rich but economically poor nations, including local communities, the opposing camp refutes this argument by raising their concerns about the extinction of species. This debate has been ongoing for a considerable number of years and the CITES Convention (both the Conference of the Parties—COP—and the Standing Committees—SC) has become the locus of a virtual battleground between participants from these two camps of belief (Stoett, 2002); Danaher, 2008; Moore, 2010; Rosenzweig & Van Weering, 2004; Sukumar, 2003). At a theoretical level, the beliefs can be categorised as follows:

- Ecocentrism: Ecocentrists opine that nature and wildlife should be preserved in as 'pristine' a condition as possible, and that the widespread utilisation of (usually select) species should be abandoned in favour of a new code of environmental ethics.

- Anthropocentrism: Anthropocentrics believe that the utilisation of species brings income and wealth to local communities, which in turn encourages these communities to preserve or save nature.
- Hierarchism: Hierarchists believe that a middle path between preservation and trade is possible. Hierarchies are the benign authorities who ensure that the various conditions for the playing of this trading game (a 'level playing field', for instance) are in place. The hierarchists do not oppose trade or calls for preservation, but rather propose a rationality on decisions which considers larger socio-economic conditions (which need not necessarily be science-based). This belief is mainly associated with power-oriented structures involved in governing CPRs.

(iv) *Basic constitutional structure:*

Before describing the relevant constitutional rules within the wildlife policy subsystem, first it is important to outline how the rules were shaped or derived based on the significant influence on how natural resources were managed by the Range States.[5] While framing the rule or control mechanism, governments had to look into the five property rights (see Ostrom, 2009b) that individuals use within a common pool resource:

1. Access—The right to enter a specific property.
2. Withdrawal—The right to harvest specific products from a resource.
3. Management—The right to transform the resource and regulate internal use patterns.
4. Exclusion—The right to decide who will have access, withdrawal or management rights and.
5. Alienation—The right to lease or sell any of the other four rights.

A well-managed forest or wildlife resource at the national level effectively balances all the above criteria through institutional rules and regulations. As mentioned earlier, in most countries, forests and wildlife are a property of the government which controls or manages the resources. In some countries, management is left to responsible private entities, who, under the basic conservation rules of the state, carry out the task of the use and conservation of wildlife and forests. In all cases, forests and wildlife are the sovereign property of the state. No external actors can directly influence their management. Now, when it comes to the governance and management of forests and wildlife, there is a great variety in the level of effectiveness, efficiency and capacity within states. There is little doubt that forests and wildlife are mostly well regulated within developed countries. However, some developing countries, although they are resource-rich, find it difficult to design and implement good management practices. At an international level, the rules of trade on fauna and flora are universally prescribed. As mentioned in Chap. 2, the main objective of the CITES Convention is to ensure that international trade in specimens

[5] Range State is a term generally used in zoogeography and conservation biology to refer to any nation that exercises jurisdiction over any part of a range which a particular species, taxon or biotope inhabits, or crosses or overflies at any time on its normal migration route.

of wild animals and plants does not threaten their survival. For this purpose, the Convention uses a permit mechanism that allows species to be listed in three appendices based on the level of their threat of extinction (see Chap. 2, Subsection 2.3.2). As CITES is a non-self-executing body, Parties are obliged to prohibit trade that contravenes the Convention. Article II, Paragraph 4, requires that: 'The Parties shall not allow trade in specimens of species included in Appendices I, II and III except in accordance with the provisions of the present Convention'. Therefore, Parties have the obligation to prohibit trade in CITES specimens whenever the Convention's conditions have not been complied with. Article VIII, Paragraph 1, supplements this general rule, requiring the State Parties to criminalise and enforce any violation of CITES prohibitions. The treaty states: 'These shall include measures: (a) to penalise trade in, or possession of, such specimens, or both; and (b) to provide for the confiscation or return to the State of export of such specimens'. However, the breach of this obligation does not itself constitute a criminal offence for the simple reason that, a decision made at CITES requires either the amendment of existing rules or the adoption of new rules by the respective national governments. Considering the fact that national laws may not change immediately after the amendments, the chances for discrepancies in the implementation of the Convention at a national level are quite high. Therefore, Article VIII, Paragraph 1, of CITES can be classified as a non-self-executing provision because it is difficult for Parties to implement it until they can adopt specific legislation for that purpose.

Therefore, the universalisation of rules is the biggest problem for CITES. It can only advocate and can never impose rules and regulations on a Party. In a world that is split by various levels of governance frameworks, universalisation of the rules is difficult to implement. For instance, CITES functions well in countries where the control or rights over the use of natural resources is strongly centralised and efficiently managed, where citizens have legal rights to use the resources only as permitted by government agencies and where this central control is popularly accepted. In such systems, the national bureaucracy is well placed to implement CITES controls effectively, as CITES laws provide an impetus to the countries' own law enforcement efforts in controlling illegal or excessive trade, thereby improving the implementation of its own policies. The CITES Convention presupposes a high degree of mutual respect for the sovereign rights of nations, and tolerance of a wide variation in approaches to conservation issues, and only interferes with matters regarding species that are listed within the Convention Appendices. On the other hand, CITES gets into trouble in situations where control over flora and fauna is neither centralised nor popularly accepted or where the state bureaucracy is weak and inefficient; here no amount of controls at the international level can effectively rectify the weaknesses of state agencies. These states are susceptible to influence from other factors, including economic interest groups. As it appears, a large majority of CITES Parties which are rich in flora and fauna have weak governance; while, at the same time, they are the primary source of wildlife products for rich countries (Fig. 3.3).

Another problem with CITES rules is that the obligations from the Convention cannot be enforced in the courts, and penalties cannot be applied for

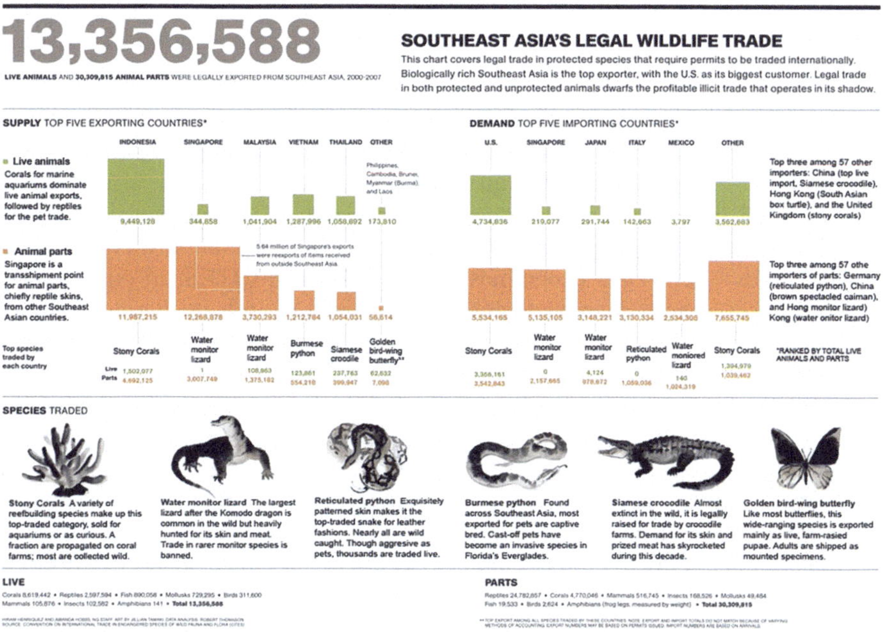

Fig. 3.3 Major exporting and importing countries of wildlife (Source: National Geographic Magazine, NGM, 2010)

non-compliance unless it is explicitly stated by the domestic legislation. This means that national laws should adhere to the primary rule system (see Chap. 2, Section 2.3.2) of the Convention. However, as it appears, the domestic governance framework in many countries is not sufficiently adequate to fulfil the requirements of CITES. The scope of this legal provision is quite broad and allows the Parties discretion on how to enforce the Convention. Some countries that are part of the Convention have still not enacted specific legislation to implement it. These countries, instead, rely on general wildlife and forest laws, and sometimes use their customs or foreign trade legislation to control trade in CITES species. These laws usually do not have the specific purpose of implementing CITES because they were enacted before CITES was signed.

External (System) Events

As mentioned in Chap. 2, CITES processes have been affected by several external system events. The formation of the CITES Convention itself was due to an external-event wave (for example, the larger environmental movement itself) against the utilitarian approach which generated steam in the late 1960s and early 1970s. After the formation of CITES, compromise approaches between trade and conservation in the early 1980s promoted a new concept called 'sustainable use' where trade and conservation were directed to coexist. These decisions had deep implications on the functioning of CITES. Some of the factors which influenced this phenomenon are briefly listed below:

(a) *Changes in socio-economic conditions and technology:* CITES came into existence during the peak period of the environmental movement, primarily due to the pressure exerted by IUCN on the world leaders by highlighting the concerns caused due to indiscriminate trade on wild flora and fauna (see also Chap. 2, Section 2.4). Since then, a strong preservationist policy has prevailed within CITES which continued until the early 80 s. The preservationist era ended in the 1980s when Asian economies gathered force in CITES. In other words, social, ethical and economic factors have played a key role in altering positions and policies related to trade in wild flora and fauna within the CITES Convention.
(b) *Changes in systemic governing coalitions resulting from the membership of countries:* As mentioned in Chap. 2, the influence of Party membership within CITES is clearly visible in the results and outcomes generated from the Convention (see Sect. 2.3 of Chap. 2) and in the Bern and Fort Lauderdale Criteria. While Parties to CITES adopted the Bern Criteria in 1976, providing more impetus to preservation values, the Fort Lauderdale Criteria (which replaced the Bern Criteria), adopted in 1994, promoted a more utilitarian approach (Sands, 2003). This shift apparently happened after powerful Asian economies joined the Convention in the 1980s—a classic example of how new Party membership can influence coalition formation and decisions made by CITES.
(c) *Changes resulting from the impact of other policy subsystems:* Although CITES operates at an international level, the success of the Convention depends on how well it is implemented at the national level by the ministries concerned, including the ministries responsible for the protection of the environment and the ministries dealing with trade. In addition, the ministries dealing with law enforcement (police, customs and border police) also play a significant role in securing compliance with the CITES Convention at the national level. The ministries of foreign affairs act at the international level and are primarily involved in negotiations and bridging bilateral partnerships. Their primary role is in supporting their national positions at the international level through the negotiation process. In other words, the wildlife–trade policy subsystem in which CITES operates is influenced by four other overlapping policy subsystems—trade, law enforcement, environment and foreign affairs. Therefore, the beliefs within the wildlife policy subsystem carry the characteristics of the actors within these other four subsystems, and any change in any of them can bring about changes in the wildlife–trade policy subsystem (see also Chap. 5, Section 5.2.1.2).
(d) *Changes in public opinion:* Though, in general, wildlife-related issues are not a major priority for the public in most parts of the world, issues influencing iconic species such as Elephant, Rhino, Tiger and Gorilla, often trigger public attention. The ivory ban in 1989 can be considered as one of the decisions that were influenced by a general public outcry.

Long-Term Coalition Opportunity Structures

What is clear from the previous section is that any change in CITES regulations requires a lengthy political and policy process.[6] This is also the case with national legislation. In some countries, laws cannot be changed easily and require a lengthy legislative process that at times even requires constitutional changes in order to conform to the regulations of a UN convention.[7] Hence, the need for consensus and openness among the coalitions is a key to implementing effective regulation. In other words, within the CITES policy context, it is difficult to achieve consensus because of an excess of veto options. Second, the system is not open, but rigid. Only states are allowed to vote and even though NGOs and business firms are now being listened to, they have no decision-making rights whatsoever.

In preparation for CITES COP meetings, CITES Parties also form coalitions. The ivory issue is one area where a distinct Party coalition can be observed. Even though there is a high level of polarisation among the various coalitions, not all decisions are put to a vote. Instead, certain decisions at CITES are based on *consensus*. The degree of consensus, of course, varies according to the *openness of the political system* of each country and the issue in context. One big challenge that arises during CITES COP meetings is that certain countries will fix their support/objection prior to the COP meetings, meaning that decisions are not based on debates and discussions within the COPs. Hence, in order to win a vote, long-term and temporary coalitions are formed. A classic case is that of China and Japan where they support each other (especially when voting on elephant at CITES) for mutual benefit at CITES and then oppose each other at other venues or policy subsystems where discussions on trade and foreign affairs are intense. The purpose of coalition formation is to advance a particular policy proposal. This can also happen in the case of capacity development efforts where NGOs and governments join forces to work on a certain project, exclusively defined for a particular purpose. These are called 'coalitions of convenience' and, depending on the circumstances, may or may not lead to long-term coalitions (see Mintrom & Vergari, 1996).

---

[6] Article XVII—Amendment of the Convention (extracted from CITES website: http://www.cites.org/eng/disc/text.php#XVII)

1. An extraordinary meeting of the Conference of the Parties shall be convened by the Secretariat on the written request of at least one-third of the Parties to consider and adopt amendments to the present Convention. Such amendments shall be adopted by a two-thirds majority of Parties present and voting. For these purposes 'Parties present and voting' means Parties present and casting an affirmative or negative vote. Parties abstaining from voting shall not be counted among the two-thirds required for adopting an amendment.

2. The text of any proposed amendment shall be communicated by the Secretariat to all Parties at least 90 days before the meeting.

3. An amendment shall enter into force for the Parties which have accepted it 60 days after two-thirds of the Parties have deposited an instrument of acceptance of the amendment with the Depositary Government. Thereafter, the amendment shall enter into force for any other Party 60 days after that Party deposits its instrument of acceptance of the amendment.

[7] See United Nations Handbook for Parliamentarians—Convention on the Rights of Persons with Disabilities—Chap. 5—*National Legislation and the Convention*.

Short-Term Constraints and Resources of Subsystem Actors

A major factor that influenced CITES, and specifically decisions on enforcement matters, was the level of financial contributions from Parties and the state of affairs within the global financial markets. With Party membership in CITES increasing, the task of the Secretariat proportionately increased. This meant that the Secretariat had to outsource some of its work to trusted partners. Financial shortfalls within the Secretariat since the start of the Convention had been considerably hampering its efforts to build capacity (hiring new staff and filling vacant positions). The most notable example could be seen during the deliberations at CITES COP14, which took place soon after the financial crisis. During the plenary session of the meeting, the CITES Secretary-General pointed out that '*anything less than a 15% increase in its budget could entail: staff cuts; a severe decrease in activities including translation services; and limited documentation being available at meetings. Any fundraising efforts would be ineffective without a stable Secretariat*' (CITES, 2007d).

3.3.2 Internal Factors Affecting the Policy Subsystem: Coalitions of Beliefs, Actors and Policy Process

The policy subsystem in which the CITES Convention operates is generally referred to as the Wildlife–Trade policy subsystem as it attends to both the protection of endangered species from unsustainable trade and at the same time promotes sustainable trading of species that are not endangered. For the same reason, the subsystem is composed of actors who are conservationists and those who are traders or actors who have a commercial interest. Apart from the core interest groups, there are government officials representing each ministry (trade, environment and foreign affairs) and several representatives from the academic community acting as knowledge brokers. At the international level, national governments or Parties to the CITES Convention alone can participate in a decision-making process. The CITES Secretariat acts as the main policy broker and facilitates the decision-making process between the Parties. It publishes position papers on the status of species and brings its views on relevant decisions to the attention of the Parties. Decision-making at the CITES level is made either by consensus or through voting. In the case of voting on species listed in Appendix I and II, a two-thirds majority alone can finalise a key decision, especially on changing the rule on up-listing or down-listing a species from its appendices. The role of NGOs at both national and international levels is that of a watchdog. NGOs alert CITES and national governments on the effects of policy decisions within a social or environmental context. Similar to any other MEA (Sewell, 2005), the actors involved in CITES function at both national and multilateral levels (see Fig. 3.4).

Due to this same reason, the coalitions of actors within the subsystem are transnational advocacy networks which exist and operate beyond the framework of a state or a Party (in MEA terms), and carry within them certain core ideological

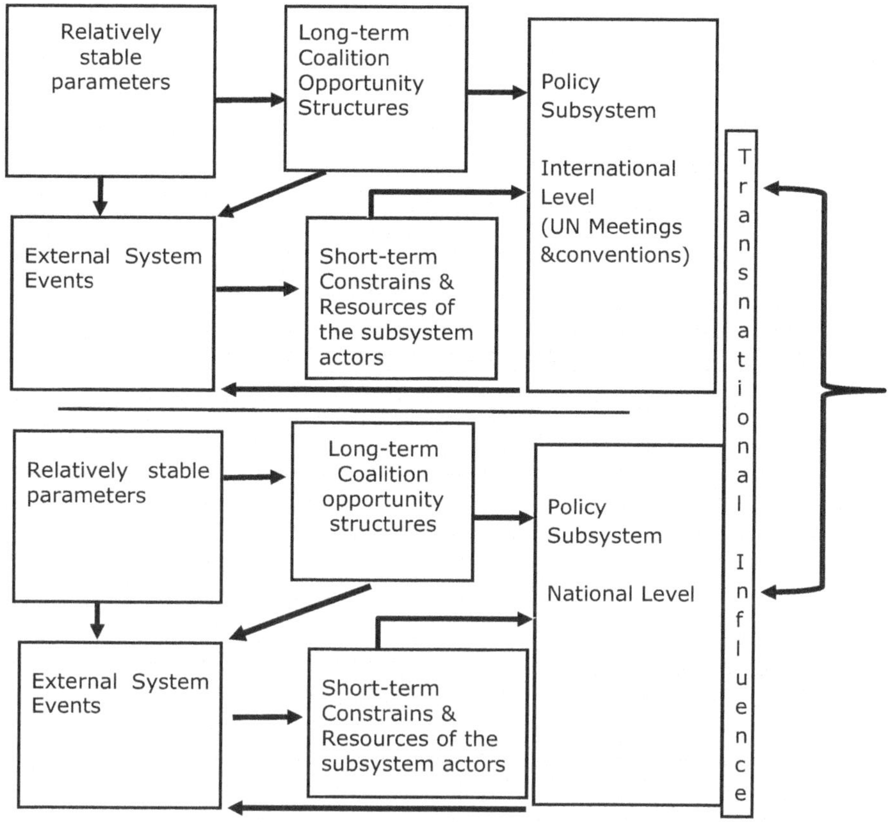

Fig. 3.4 Parallel international and national policy subsystems

perspectives or *Core Beliefs (CB)* on the constructions of 'nature', 'economy' and 'livelihood' that apply to multiple policy domains. *Policy beliefs (PB)* translate these core beliefs into a position supporting either conservation or utilitarianism. Coalitions are then formed among the actors holding similar policy beliefs. Similar to other Multilateral Environmental Agreements, a linkage exists within CITES processes between advocacy networks within national and international policy subsystems (see Fig. 3.4). The intention of this link is to influence the behaviour of states internationally in order to gain acceptance of the policies or beliefs of the network. For example, when the links between state and domestic actors are severed, domestic NGOs may directly seek international allies in order to apply pressure on their own states from outside (Keck & Sikkink, 1999) or at the CITES Conference of Parties itself. Keck and Sikkink (1999) refer to this as the 'boomerang' pattern of influence—a characteristic of transnational networks where the target of their activity is to change a state's behaviour. In policy science, this phenomenon is usually referred to as 'venue shopping'—this is where policy actors will prefer venues (policymaking or legislative or judicial procedures) where they have the best chance of getting a favourable decision on their cause. If this fails several times, they find—or create—other venues and try again.

In summary, it is now clear how CITES processes fit well within the ACF general framework. The section has also described the transnational nature of the actors influencing decisions on CITES. In the next section, an example is elaborated through a case study of big-leaf mahogany listing, to provide a more detailed picture of the interactions of actors within the wildlife–trade policy subsystem.

3.4 Case Study: Big-Leaf Mahogany and Transnational Coalitions

Big-leaf mahogany (*Swietenia macrophylla*) is considered to be one of the most valuable tropical timber species and is native to the Central and South American region. During the early 1990s, it fetched up to $1600 per cubic metre, representing big business in the United States, an importer of more than 60% of all the mahogany exported from Latin America (Bohlen & Sandalow, 2002). Big-leaf mahogany is one of the most exploited species, with most of it originating from Peru and the Brazilian Amazon (Roozen, 1998).

In the early 1990s, following alerts from investigative conservation NGOs, the Bush administration decided to control the import of unsustainable big-leaf mahogany from South American countries to the United States and other regions. In 1992, the United States submitted a proposal along with Costa Rica at the eighth meeting of the Conference of the Parties (COP8, Kyoto) to include this species in CITES Appendix II (CITES, 2014). The argument of the proponents was that the extreme exploitation of the timber in the Range States should be controlled by regulating trade through quotas and export permits. However, their proposal was rejected by the CITES Parties, mainly producer and consumer countries backed by international timber traders and lobbyists. There were several reasons for this rejection. First, there was no scientific evidence to prove that big-leaf mahogany was threatened, nor that it was a factor in Amazonian deforestation (Roozen, 1998). Second, the timber lobbyists in the USA and in big-leaf mahogany-producing countries were powerful enough to gain the confidence of several governments in supporting their position at both national and international levels. The ultimate outcome was that the US–Costa Rica proposal to list the species in Appendix II failed.

After the first defeat, conservation NGOs, including Greenpeace, which had earlier worked with the US government to list the species, lobbied the UK and US governments to put pressure on 'weak' states to control illegal trade. At the same time, Dutch forestry scientists engaged with their government to raise the issue of the threat to South American big-leaf mahogany (Bonner, 1994). As a result of all these actions, at CITES COP9 (Fort Lauderdale, 1994), the Netherlands government submitted a proposal similar to the one submitted at the previous conference by the United States and Costa Rica (CITES, 2014). This time the difference was that NGOs had already documented evidence of illegal timber trade. As the CITES COP meeting was based in the United States, conservation NGOs could use the advantage of proximity to bring the case before the highest authorities in the USA. During the Fort Lauderdale meeting, all the circumstances favoured a listing of the Dutch proposal in Appendix II. However, it got defeated, being just six votes short of the two-thirds majority needed for adoption (CITES, 2003). The coalition of timber traders had won again despite the heavy pressure applied by conservation groups and scientists.

By 1995, scientific evidence of the threat of big-leaf mahogany extinction became clearer (Roozen, 1998). Public awareness increased and small demonstrations calling for the protection of the species were organised in the South American Range States. In an attempt to save its species, in the same year, Costa Rica included big-leaf mahogany in Appendix III (Appendix III listing does not require a vote; countries can unilaterally decide on this option), restricting the listing to the species populations in the Americas. The impact of Costa Rica's action was not confined to itself, indeed it was felt across major big-leaf mahogany-producer states ranging from the southernmost part of the species range in Bolivia and Peru to its northern limits in Mexico, and by consumer states in North America and Europe.

There was a third attempt to include big-leaf mahogany in Appendix II when a proposal was made by Bolivia and the United States at the CITES COP10 meeting in Harare in 1997 (CITES, 2014). This attempt faced another defeat when it was put to the vote. However, on this occasion there was a slight change as the situation had raised some global concern regarding the future of the species. The scientific facts that NGOs and scientists revealed during the meeting led to a compromise deal where Brazil proposed the creation of a working group to study the status of the species under the auspices of the Amazon Pact Treaty, incorporating all Range States, importing countries and expert organisations (CITES, 2014). The intention of this Brazilian proposal was to develop recommendations within 18 months on conservation measures for big-leaf mahogany. The Conference of Parties agreed to Brazil's proposal and decided at that meeting to establish a working group, comprising the Range States and importing countries, with the task of examining the conservation status of big-leaf mahogany and making recommendations to ensure sustainable international trade. After that meeting, Bolivia, Mexico and Brazil decided to include their mahogany populations in Appendix III in 1998.

At the CITES COP11 meeting in 2000, the delegation of Brazil presented a report, summarising the results of the working group from June 1998. Brazil outlined its actions in relation to big-leaf mahogany including: a decrease in exports since 1990; legal action aimed at reducing the exploitation of the species for the period 1996–2000; adoption of a licensing procedure following Appendix III listing; adoption in 1999 of a National Forest Programme, incorporating sustainable forest development; and the finalisation of a project for the sustainable production of big-leaf mahogany timber (CITES, 2000; CITES, 2000c). After CITES COP11, Colombia and Peru also listed the species in Appendix III. However, Appendix-III implementation remained problematic, undermining the effectiveness of this listing in reducing illegal trade (CITES, 2003). Evidence of the problems in controlling illegal trade was provided in the national reports presented at the 2001 meeting of the CITES Mahogany Working Group in Bolivia, as well as in a TRAFFIC review of CITES implementation undertaken at the request of the CITES Secretariat (CITES, 2003). Concerns regarding illegal harvests in Brazil prompted it to suspend all harvest authorisations. Reports also emerged from Peru of illegal logging in protected areas, including Indigenous reserves. Intelligence that illegally logged timber was being exported prompted seizures of large quantities of big-leaf mahogany in the United States and Europe in 2001 (CITES, 2003). The Appendix-III listing provided a basis for these seizures. In response to information received from Brazil's CITES Management Authority, some importing Parties questioned whether the timber had been obtained in accordance with Brazil's laws for the protection of fauna and flora, as required under Appendix III.

As the momentum was building up to protect big-leaf mahogany, the timber and furniture industry started lobbying the Bush administration against any actions that might limit their supply, while anti-trade groups were convincing Nicaragua and Guatemala to propose the listing of their species at the CITES meeting in 2002. As the CITES meeting was getting closer, the trade lobbyists had almost won the support of the US government, and the US position regarding the proposal on listing the species was not positive. Brazil and Peru, who had earlier flagged the issue of listing the species within Appendix III, dramatically reneged on up-listing it to Appendix II, calling instead for the Big-leaf Mahogany Working Group to continue, supported by the International Wood Products Association. As this was a turnaround from all the earlier positions of the USA and Brazil, the policy subsystem was becoming more complex. Both traders and conservation NGOs lobbied intensively to secure their positions. However, a scathing attack that was published worldwide in the *International Herald Tribune* on the morning of the CITES vote reversed the US position. The op-ed report (Bohlen & Sandalow, 2002), headlined 'Bush Policy Sells Amazon Treasure Down the River', was authored by Curtis Bohlen and David Sandalow, who had served as Assistant Secretaries of State for the Environment under (respectively) the George H.W. Bush and Clinton administrations. This article changed the voting game, and, in 2002, the Parties accepted the proposal from Nicaragua and Guatemala to list big-leaf mahogany in Appendix II, effective from 15 November 2003—eight years after the Appendix-III listing first came into effect. The voting amongst the Parties was also quite tight, requiring two-thirds majority excluding absentees (see Article XV(b) of the CITES Convention).[8] The voting concluded with 68 *'for'* and 30 *'against'* with several *'absentees'*. There was another reason for the proposal to go through. During that time, the CITES position[9] on

[8]Article XV(b) of the CITES Convention: Amendments shall be adopted by a two-thirds majority of Parties present and voting. For these purposes, 'Parties present and voting' means Parties present and casting an affirmative or negative vote. Parties abstaining from voting shall not be counted among the two-thirds required for adopting an amendment.

[9]CITES position: The Secretariat believes that the fundamental problems related to the conservation and management of, and trade in, this species cannot be resolved by its inclusion in Appendix III, and recommends that the Range States seriously consider supporting an Appendix-II listing.

listing the species was also favourable, with a recommendation for the Range States to seriously consider supporting an Appendix-II listing.

For more than a decade, conservationists had sought to protect big-leaf mahogany under the Convention on International Trade in Endangered Species (CITES). Each time, however, they were thwarted by the trade-supporting coalitions. This example of the listing of mahogany highlights the effects of long-term coalitions and policy beliefs at an international level on reaching a decision within a policy subsystem. Here, two coalitions existed at the international level: one with a strong ecocentric approach (Coalition A) supporting the listing and protection of big-leaf mahogany; the other constituting timber and furniture traders (Coalition B) fighting against the listing. Since only governments can vote at CITES COP meetings, it was important for both coalitions to convince their governments to support their respective positions. For this reason, the lobbying efforts were more intense at the national level. At the international level, the interest was first generated by prominent international NGOs who had alerted their governments to the extent of the exploitation of the species. Therefore, constant interactions between national and global policy subsystems took place before a decision at the national and international level could take effect.

Another important aspect in the process was the role of policy brokers within the international policy subsystem. As can be seen, CITES (the main policy broker) was not supportive of the listing of the species until 2002 as there was no scientific or other supportive evidence to show that the trade was detrimental to the survival of the species. But by 2000, and as scientific evidence emerged, the CITES Secretariat played a key role in favouring the big-leaf mahogany decision. First, it advised the Parties to form a working group on the species and provide scientific evidence concerning the status of the species and the relevant legislation. Second, CITES declared its own position, recommending that the Range States seriously consider supporting an Appendix-II listing. These factors finally led to the listing of mahogany during CITES COP12.

While this was the state of affairs at the international level, actions at the national level had considerable influence on these international-level decisions. For example, lobbying within the Range States, especially in Brazil and Peru, was crucial for the timber traders. Within the USA, there was a strong move by the conservation coalition to defeat the trade coalition. The article in the *International Herald Tribune* on the morning of the vote was a key factor which led to a change in position by the USA, and shows how coalitions use media to influence decisions. The above case describes the roles and behaviour of coalitions within a policy subsystem. It also shows how actors use resources to advance their policy positions. One lesson that can be learnt from the above case is that CITES policy processes should not only be dealt with at an international level but at a national level as well, since most of its decisions are impacted by actions at the national level.

3.5 Conclusion

In this chapter, the CITES policy process is illuminated through the lens of the Advocacy Coalition Framework (ACF), where policy changes are outlined based on the beliefs and behaviours of actors or groups of actors working within the policy domain. As mentioned in the introduction, the analysis of beliefs within the wildlife policy subsystem is specifically important as no study has conceptualised the beliefs and policy process within CITES over the course of time (during a time span of more than a decade), and of the role that coalitions and other external factors play in influencing the wildlife–trade policy subsystem.

The analysis also shows that ACF is a useful theoretical framework for uncovering many of the complexities in the wildlife–trade policy subsystem, as could be seen when it was applied to the CITES decision-making process in listing big-leaf mahogany. The case study shows that Parties will use scientific evidence to support their position but refute its validity when it does not. It is important to research this aspect further, especially as to whether such a process applies in the case of WEMS. Earlier studies (Li, 2007) have shown how international organisations, lacking democratic legitimation, have to rely on national member states for the implementation of their policies and on expertise and evidence to convince these member states to qualify, or sometimes go against, their national interests. In most cases, if evidence and/or expertise is lacking, they have no basis to work from. This means, a successful application of ACF requires significant knowledge of the stable and dynamic factors influencing a policy subsystem and a thorough knowledge of the advocacy coalitions at both national and international levels. It also requires significant understanding of the behavioural patterns of the policy broker during times of complex decision-making processes.

As this book looks into the complexities involved in the development of WEMS, this chapter has indicated the importance of understanding the beliefs and interactions of coalitions within a policy subsystem and especially within the context of WEMS development. As a step towards the application of ACF within the WEMS context, in the next chapter, the stakeholder's beliefs in the wildlife policy subsystem in Asia and Africa are first examined.

Chapter 4
Stakeholder Policy Beliefs in the Wildlife Policy Subsystem in Asia and Africa

4.1 Introduction

In the previous chapter, the CITES policy process was examined through the theoretical lens of advocacy coalition framework. Further, using an example on decisions related to big-leaf mahogany at CITES COP meetings, the influence of external factors on the policy subsystem and the interaction of the various coalitions and the role of the policy broker in the decision-making process at CITES was outlined. In the same vein, in this chapter, within the context of WEMS development, I empirically examine the various policy beliefs of the stakeholders involved in WEMS policy process in Asia and Africa.

Before moving to the details of the empirical study, using a short narrative, I will first describe the relevance of the study and why it is important to understand the policy beliefs within the context of WEMS development and implementation.

On the third of June in 2005, a busy fifth series of International Conference on Environmental Crime was hosted by INTERPOL in its headquarters in Lyon. The aim of the conference was to enhance international efforts to fight environmental crimes—then considered as one of the fastest growing crime areas. Several senior law enforcement officials and experts from 46 countries met at the General Secretariat in Lyon to discuss a variety of issues, including links between environmental crimes and organised crime and bioterrorism, investigative techniques, regional and global trends, and capacity building among international organisations. By the end of the conference, an INTERPOL specialised officer called out, 'We expect everyone to walk away from this conference with a better understanding of the seriousness of environmental crimes and the absolute necessity of more global cooperation to fight it'.

There was nothing new about the officer's message. He, like most of his fellow enforcement officials at CITES, WCO and INTERPOL, used the word 'cooperation' in tandem with the efforts needed to combat a growing crime area. But there

R. Chandran, *The Wildlife Enforcement Monitoring System*, Professional Practice in Governance and Public Organizations,
https://doi.org/10.1007/978-3-031-97084-9_4

was something else special about the meeting—INTERPOL had decided to appoint its first-ever officer specialised in wildlife crimes following a donation of US$300,000 (spread over 3 years) from three non-governmental organisations—the Bosack and Kruger Foundation, Safari Club International and the International Fund for Animal Welfare (IFAW).

A few months later, in October 2005, quite far away from Lyon UNU researchers and civil society members were applying the INTERPOL officer's word—'cooperation'—to bring in a mechanism for enhancing enforcement information-sharing on wildlife crime through the WEMS initiative (detailed in Chap. 2). Interestingly, both INTERPOL and UNU had IFAW as a common donor funding their respective wildlife crime projects. While UNU was attempting to complement the efforts of INTERPOL and CITES, hardly did it know that its work would lead to controversy and inter-agency conflict.

As mentioned earlier in Chap. 2, CITES objected to the role accorded to the NGOs as information collectors on the WEMS project, although it had no objection to the NGOs supporting the capacity development efforts of INTERPOL.

4.2 Stakeholders' Beliefs and Interaction Within the Wildlife-Trade Policy Subsystem

The history of CITES convention reveals that this was not the first time CITES was expressing concerns about NGO (other than TRAFFIC and WWF) involvement in enforcement information-collection process. It should be noted that CITES had no objection in NGOs supporting funding to INTERPOL for capacity development. The concern arises when NGOs are in custody of enforcement-related information. One may ask, why should such an objection be raised when CITES itself is preaching to national governments in sharing data with its Secretariat?

The empirical study of this research starts at this point. There are several literatures (see Duffy, 2000, 2013; Gomar & Stringer, 2011; Neumann, 1995) that delineate NGOs as political actors. They are representatives of core beliefs which are not visible to naked eyes and naked minds. For a layman, they are activists fighting for a cause. But for a policymaker, they are subjects with 'power' and 'knowledge' that can change the policy dynamics.

The history of CITES (see Chap. 2) describes the above dynamics in a more vivid manner where CITES has been accused by the various factions on the way it operates in terms of preventing and allowing small amounts of regulated international trade (Moore, 2010).

As an international policy broker, the role of CITES Secretariat is to facilitate decision-making, primarily between the Parties. It at times takes the role of a 'soccer referee', running around to see that the game of international wildlife trade is played fairly.

But bringing fairness is always a tough job. The reason is: as it happens within any multilateral agreements, at CITES meetings, actors within the wildlife-trade subsystem aggregate into a number of advocacy coalitions (Sabatier & Jenkins-Smith, 1993) that are composed of stakeholders from various organisations who share a set of policy beliefs (see Sect. 3.3) and who often act in concert. This means, coalitions (whoever they are) always look forward to winning their position rather than finding a solution to a problem. A solution emerges when the stronger coalition wins and then arrives at a decision. It doesn't matter if the decision is right or wrong, what matters is more about having a decision.

If coalition politics is one of the reasons behind the objection of WEMS, it is first important to understand the nature of the coalitions and how they operate. Hence, the primary focus of this chapter will be on the following research question:

What are the policy beliefs within the Wildlife Policy Subsystem in Asia and Africa that influenced the policymaking regarding WEMS in Asia and Africa?

To answer the above question, a detailed qualitative study using Q methodology was carried out in Asia and Africa involving all actors within the wildlife policy subsystem (including NGOs who had participated in the WEMS initiative). The detailed process of the empirical study, conceptualised using cultural theory, is described in the next section.

4.3 Conceptual Framework of the Study

The roles and behaviour of actors using (transboundary) decision-support systems like WEMS can be conceptualised by relying on social science theories of technology. Wildavsky (1987), for instance, mentions that adoption of a technology depends on the preferences of an actor's normative choice. When individuals make important decisions, these choices are simultaneously choices of shared values within the limiting patterns of social practices (Wildavsky, 1987). In other words, understanding the actors' perceptions requires understanding their cultural bias and on how normative beliefs affect technology-relevant policymaking processes. Cultural theory (Douglas, 1978) has been put to use in understanding similar policy debates on environmental problems and risks through its four myths of nature (Douglas & Wildavsky, 1983; Thompson, Ellis, & Wildavsky, 1990). Cultural theory outlines how people, despite severe time pressure and lack of in-depth expertise, quickly figure out which policy ideas and experts are the most credible ones (Gastil & Levine, 2011; Hoppe, 2009).

The process of using cultural theory as a tool in policy analysis and how it can be applied in understanding the beliefs of actors within the wildlife policy system is outlined in the following section.

4.3.1 Cultural Theory as a Tool to Understand the Social Relationships in Policy System

Cultural theory furnishes a map with four ideal types of social relationships or solidarities: individualism, hierarchism, egalitarianism and fatalism (Thompson et al., 1990). Corresponding to these four types of social relationships are cultural biases, sets of shared values and beliefs (Thompson et al., 1990). These are judgements of value which function as justifications for specific organisational structures. Each ideal type of social relationship is characterised by its own typical set of beliefs, a cognitive and moral bias that contributes to reflexivity in the social organisation. In cultural theory, knowledge about nature is not the exclusive preserve of any particular domain of society—science, bureaucracy, civil society or the market. Each domain engages in the production of its own knowledge, ideas, beliefs and meanings, and knowledge about nature and political order is 'co-produced through a common social project that shores up the legitimacy of each' (Jasanoff & Wynne, 1998).

4.3.2 Cultural Theory as a Tool to Understand the Policy Process

Hoppe (2007) describes several ways of how cultural theory can be applied as a tool to enrich policy analysis. A first application is to map the belief systems (Sabatier & Jenkins-Smith, 1993) of the protagonists and antagonists within a policy domain. According to Sabatier (2007), any policy domain, such as wildlife conservation, is made up of multiple actors who, motivated by their beliefs, structure their relationships into competing coalitions and try to influence policy by utilising multiple resources and venues.

The policy analyst can use cultural theory's four ideal types of social relationships—hierarchism, individualism, egalitarianism and fatalism—as a compass in finding his/her bearings in the ideals espoused in the belief systems. To develop the cultural theory-compass into a typology of discourses that also covers policy core beliefs and secondary beliefs, considerable substantive familiarity and interactional expertise (Collins & Evans, 2002) with the particulars of discourses in a particular policy domain is necessary. Especially policy core and secondary beliefs have to be based upon familiarity with the policy field and intensive interpretive analyses of the discourses within the policy domain in question. On this front, my familiarity with the policy domain stood as a distinct advantage in analysing policy beliefs within the wildlife policy subsystem.

As the primary focus of this chapter is on identifying the various beliefs within the wildlife policy subsystem in Asia (where WEMS was proposed to be implemented) and in Africa (where WEMS was adopted), the following sections will outline the research design that was contemplated to extract the beliefs of the stakeholders in Asia and Africa.

4.4 Research Design: Using Q-Method to Map Policy Belief Systems

In order to elicit the ways of thinking of wildlife conservation actors, the first step was to identify a suitable research design that can be used to analyse the variety of opinions, perceptions and attitudes that reflect the belief set of an individual or group of individuals in a valid and reliable way. Q Methodology (Brown, 1996; Stephenson, 1954) was identified as the appropriate research tool as it is widely used in many subject areas, including environmental policy (see Table 4.1).

Q methodology comprises four steps: (1) a selection of statements relevant to the topic; (2) administering the Q survey to participants; (3) multivariate statistical analysis to identify groups with similar belief patterns evidenced in similar statement rankings and, finally, (4) qualitative analysis of the findings resulting in a systematic overview of the perspectives prevalent among the participants. It is important to stress that the Q methodology does not result in a statistically representative description of the prevalence of belief sets. Its sole aim is to uncover the variety of beliefs among a group of participants.

Table 4.1 List of some applications of Q methodology relevant to nature policy

Authors	Aim of Study
Cairns (2012)	*Understanding science in conservation: A Q-method approach on the Galapagos Islands.*
Rastogi, Hickey, Badola and Hussain (2013)	*Diverging viewpoints on tiger conservation: A Q-method study and survey of conservation professionals in India.*
Mattson, Byrd, Rutherford, Brown and Clark (2006)	*To identify perspectives in large carnivore conservation in northern United States.*
Sickler et al. (2006)	*To determine potential visitor's social perspectives about dolphin intelligence, and how these beliefs might influence acceptance of scientific information.*
Salazar and Alper (2002)	*To understand the perspectives on democracy and social justice in British Columbia's environmental movement.*
Clarke (2002)	*To understand sustainable development in the context of other emergent environmental perspectives.*
Hoppe (2009)	*To analyse expert advice and discourses of policy workers on their jobs.*
Wolsink (2004)	*To understand policy beliefs in spatial decisions.*
Neff (2011)	*To assess what ecologists see as important research in their discipline and how they arrive at their priorities.*

4.4.1 Q-Method Design

The Q-method design used in this study is followed from Neff (2011). The various steps carried out are described in Fig. 4.1 and further enumerated below:

The first step for conducting the empirical study was to derive a large population of statements from reputable academic journals and grey literature that captured the various discourses and perspectives in wildlife conservation ('the concourse'). From the long list of statements of this concourse, a sub-concourse was created which captured the core beliefs (CB), policy beliefs (PB) and the secondary beliefs (SB) of the policy actors. In a further step, these were grouped according to cultural theory's ideal-typical biases and the major stakeholder groups in the policy domain. The final set of Q statements, or the Q-sample, is a selection from the sub-concourse which contains only clearly defined statements that systematically capture (Q samples can be small as long as it covers a broad range of opinions within a policy subsystem; see Brown, 1996) all the existing perspectives used by the wildlife policy system's major stakeholders. All statements used in the survey were extracted from the discourses of real-life policy actors.

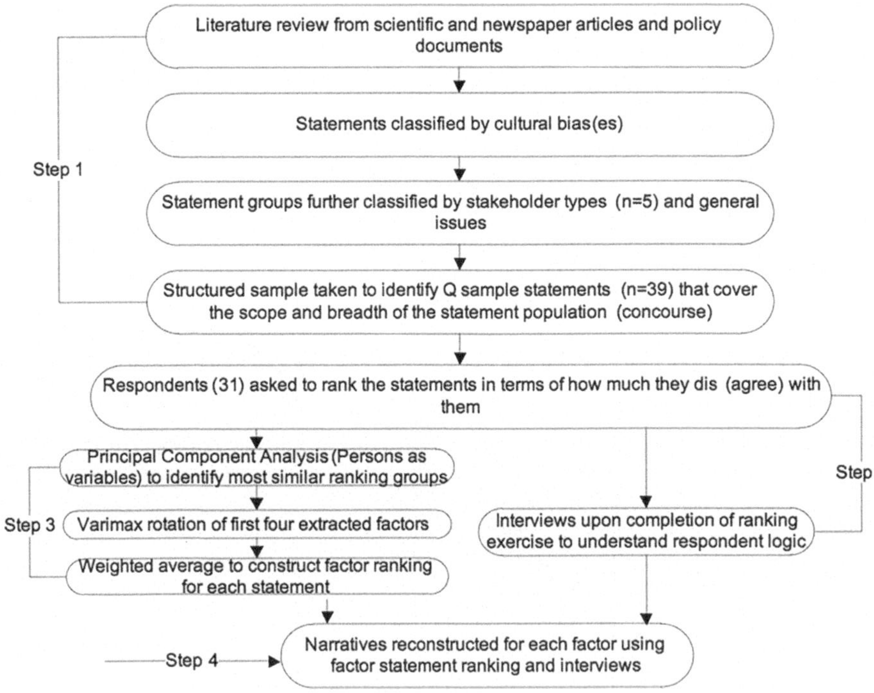

Fig. 4.1 Flowchart representing the steps and concomitant research methods used in Q-method design for this research project (cf. Neff, 2011)

Table 4.2 Chart describing the core beliefs, policy beliefs and secondary beliefs of each stakeholder within a wildlife-trade policy subsystem

Groups	Core Belief	Policy Belief	Secondary Belief
Government	Hierarchical	Rule based	Legal structures and 'cleared' (approved) information in vertical accountability structures
NGOs	Egalitarian	Rights based	Transparency and public records in horizontal accountability structures
Traders	Individualist	Profit based	Financial instruments
Local community	Fatalist	Fate based	Animism, mysticism
Academics	Hybrid	Science based	Theory-relevant information and facts

From the literatures, a set of statements (n = 296) that were deemed rich enough to cover the entire concourse[1] on wildlife conservation was selected. Statements that were specific to the regions in Asia and sub-Saharan Africa were prioritised during the selection process. The original compilation ran an entire spectrum of perspectives from the local or community level defining the rights of the local community (Guha, 1999; MOEF, 2005; Sato, 2003; Vandergeest, 1996) to the perspectives of government officials, traders and researchers in Asia and East Africa, including their views on international trade and global conservation issues (Danaher, 2008; Hutton & Dickson, 2000; Moore, 2010; Nijman, 2010; Swinbanks, 1987). A second source of statements was obtained from the news archives of the BBC, Reuters and the *Guardian*. The third source was the set of documents generated in several CITES meetings of the Conference of Parties. From the three sources, statements were selected that describe a *sub-concourse* of possible stakeholder types in the wildlife conservation policy subsystem. A chart (Table 4.2) was then developed in accordance with cultural theory to bring out the patterns of statements that capture the perspectives of major stakeholders in the policy system, including the academic community (a subset was added on the basis of empirical knowledge and familiarity of the field).

By selecting two statements for each category, a total of thirty (2 x 5 x 3) Q statements were derived. Additionally, nine more statements on general policy issues were added to identify any other beliefs not described by the literature but based on the characteristics of the policy domain. The final Q-sample comprises 39 statements, randomly numbered in Table 4.3. As can be observed from Table 4.3, the only variation in the Q statements for Asia and Africa was in #28, #29 and # 31.

According to Brown (1996), an important notion behind the Q methodology is that only a limited number of distinct viewpoints exist on any topic. Any

[1] We used similar statements for Asia and Africa, different only in their relevance to the geographic context.

Table 4.3 Statements used in the Q samples in Asia and Africa

	Statements for Asia	Statements for Africa
1	Poachers should be shot when spotted within a protected forest area	Poachers should be shot when spotted within a protected forest area
2	Wildlife should be protected as any kind of its use will lead to species extinction	Wildlife should be protected as any kind of its use will lead to species extinction
3	Wildlife is nature's gift which should not be substituted for economic subsistence	Wildlife is nature's gift which should not be substituted for economic subsistence
4	Wildlife trade and hunting should be banned	Wildlife trade and hunting should be banned
5	Conservation NGOs should be provided legal rights to stop wildlife crime	Conservation NGOs should be provided legal rights to stop wildlife crime
6	Conservation NGOs should be the custodian of wildlife crime database	Conservation NGOs should be the custodian of wildlife crime database
7	Wildlife policies should NOT allow options to cull endangered species when it causes crop and human damage	Wildlife policies should NOT allow options to cull endangered species when it causes crop and human damage
8	Government prioritise wildlife conservation over wildlife trade	Government prioritise wildlife conservation over wildlife trade
9	Local communities should be provided legal rights to hunt and sell wildlife for sustenance	Local communities should be provided legal rights to hunt and sell wildlife for sustenance
10	Wildlife policies should provide options to cull endangered species when it causes crop and human damage	Wildlife policies should provide options to cull endangered species when it causes crop and human damage
11	Policies supporting sustainable use of wildlife is good to maintain a healthy wildlife population	Policies supporting sustainable use of wildlife is good to maintain a healthy wildlife population
12	Protection of wildlife can only be in zoos and museums	Protection of wildlife can only be in zoos and museums
13	Bush meat should be allowed as food for poor	Bush meat should be allowed as food for poor
14	Wildlife is an economic commodity and should be utilised	Wildlife is an economic commodity and should be utilised
15	Wildlife population can naturally revive and hence does not need significant restrictions on its consumption	Wildlife population can naturally revive and hence does not need significant restrictions on its consumption
16	Wildlife commodity should be listed in financial stock markets so it fetches uniform and better values	Wildlife commodity should be listed in financial stock markets so it fetches uniform and better values
17	Science should determine the core policy on whether wildlife should be traded or not	Science should determine the core policy on whether wildlife should be traded or not
18	Scientific information should determine whether hunting should be allowed or not	Scientific information should determine whether hunting should be allowed or not
19	Science should determine how nature should be managed	Science should determine how nature should be managed

(continued)

Table 4.3 (continued)

	Statements for Asia	Statements for Africa
20	Wildlife cannot be conserved without addressing the rights of the forest community	Wildlife cannot be conserved without addressing the rights of the forest community
21	Species listing of CITES appendices should be based on scientific information and not voting by member states	Species listing of CITES appendices should be based on scientific information and not voting by member states
22	As long as utilisation of wildlife is regulated, wildlife remains protected	As long as utilisation of wildlife is regulated, wildlife remains protected
23	Reputed academic institutions should manage wildlife crime databases	Reputed academic institutions should manage wildlife crime databases
24	Only key wildlife species need to be protected as some can be traded	Only key wildlife species need to be protected as some can be traded
25	Marine species are losing out as protection of terrestrial mammals is getting precedence.	Marine species are losing out as protection of terrestrial mammals is getting precedence.
26	At certain point of CITES negotiation, trade-off of certain key marine species is considered better than allowing Tiger farms and ivory trade	At certain point of CITES negotiation, trade-off of certain key marine species is considered better than allowing Tiger farms and ivory trade
27	Bird species are losing out as protection of terrestrial mammals is getting precedence	Bird species are losing out as protection of terrestrial mammals is getting precedence
#28	Tuna should not be listed in CITES appendix as it is used as a major food product	African fish species should not be listed in CITES appendix as it is used as a major food product
#29	Tiger conservation need more financial support than any other endangered species	Elephant conservation need more financial support than any other endangered species
30	Species-specific databases are most important than a general illegal trade database	Species-specific databases are most important than a general illegal trade database
#31	Tiger and Elephant crime databases will be enough for monitoring effective compliance of traded species	Elephant databases managed by CITES are the most trusted databases
32	Good funding for enforcement activities brings effective compliance of CITES	Good funding for enforcement activities brings effective compliance of CITES
33	The national wildlife crime information systems should be managed by a private IT company	The national wildlife crime information systems should be managed by a private IT company
34	Professional IT consultants alone can bring out a good wildlife crime database	Professional IT consultants alone can bring out a good wildlife crime database
35	Governments should take the lead in addressing all issues with regard to enforcement and compliance of CITES	Governments should take the lead in addressing all issues with regard to enforcement and compliance of CITES
36	Governments alone are the custodian of illegal wildlife trade information	Governments alone are the custodian of illegal wildlife trade information

(continued)

Table 4.3 (continued)

	Statements for Asia	Statements for Africa
37	The UNEP-CITES secretariat should emphasise stricter sanctions against non-complying states	The UNEP-CITES secretariat should emphasise stricter sanctions against non-complying states
38	Harvest-related measures and trade-related measures should be used in tandem, to ensure the successful management of natural resources	Harvest-related measures and trade-related measures should be used in tandem, to ensure the successful management of natural resources
39	Conservation practices should also focus on income generation so as to achieve sustainability of such practices	Conservation practices should also focus on income generation so as to achieve sustainability of such practices

well-structured Q-sample, containing the wide range of existing opinions on the topic, will reveal these perspectives.

The second step was to administer the Q-sample to a set of participants (P-sample). In Q methodology any sample of a little over n = 20 is usually sufficient. The participants (anonymous but for their function) who make up the list were all involved in wildlife management issues at the national and global levels. Their professional functionality is listed in the first column of Table 4.6 (for Asia) and Table 4.9 (for Africa).

Jointly, they can be considered as the community of policy experts in wildlife conservation within their own countries. All the participants were asked to rank the 39 statements of the Q-sample in terms of how much they agreed or disagreed with them. The Q-sort table was designed as a forced Gaussian distribution with seven categories, running from 'most disagree' (−3), neutral (0) to 'most agree' (+3). The number of statements per category was - 2, 5, 8, 9, 8, 5 and 2. Upon completion of ranking the Q statements, participants were qualitatively interviewed. These were semi-structured interviews focused on uncovering the participants' reasoning and logic in sorting the Q statements. The participants were also interviewed on their familiarity, knowledge and their concept of WEMS design. The interview summaries were manually coded (Miles & Huberman, 1999) and related to the results of the statistical analysis and the belief classification matrix given in Table 4.2.

In a third step (Step 3 in Fig. 4.1), Q-sorts were statistically analysed using the PQ-Method Software (Schmolck, 2014). Principal Component Analysis (PCA), with persons as variables, was used to discover clusters (or groups) of participants with the most similar statement rankings. The statistical procedure derived 8 factors, classified based on the *eigenvalues.* Using Scree test (Cattell, 1966), four factors (optimal factors) were chosen for performing a VARIMAX rotation.

As Brown (1996) advises, weighted averaging of the scores was used to calculate each factor's (group's) average rating for each statement (see Table 4.5 for Asia and Table 4.8 for Africa). Based on these premises, it was made clear that there were statements from all the three levels of belief systems: core beliefs (CB), policy core beliefs (PCB) and secondary beliefs (SB). In addition, each participant's ranking was transformed into a factor loading, signalling the degree to which an individual's ranking corresponds to the averaged rankings of the four principal factors (see Table 4.6 for Asia, and 4.9 for Africa).

The *fourth and final step* was a comparison of different clusters (groups) involving teasing out of underlying rationales per factor and their differences. The comparison was done by looking at the (statistically) defining statement per factor and the 'most agree' and 'most disagree' statements of the high-loaders on each of the four factors.[2] By an extensive iterative process of triangulation between the ranking data of individual but 'typical' participants, averaged cluster rankings and the qualitative interview data, 'narratives' (M.D. Jones & McBeth, 2010) were prepared that correspond to the views of each cluster on wildlife management and policy. These findings are reported in the results section.

4.5 Results

The presentation of the results is based on the Webler (2009) and M.D. Jones and McBeth (2010) method, who present the belief clusters by a descriptive label followed by an accompanying narrative. Each narrative describes the policy beliefs of a coalition (as the factors bringing together views of the actors representing a particular belief system) with respect to conservation, enforcement, compliance and information-sharing. The analysed data gave clear evidence on the existence of four different policy belief systems which remained similar (if not the same) in Asia and Africa. In the next section, separate analyses of the beliefs are carried out for Asia and Africa and, in the following section, comparative analyses of the beliefs are done.

4.5.1 Analysing Policy Beliefs in Wildlife Management in Asia

For the purpose of identifying the beliefs within the wildlife policy subsystem in Asia, Q-sort was carried out in India, Thailand and Japan. The reasons for selecting these 3 countries were these:

a) The civil society network which participated in the design process of WEMS was from these three countries.
b) WEMS was also planned to be implemented in these countries after its development.

---

[2] # represents the statement and the +/− value represents ranking of the particular statement in the factor.

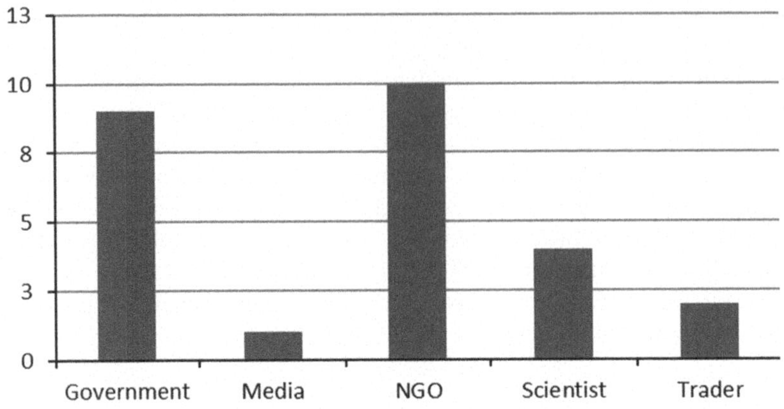

Fig. 4.2 Category of participants in the Asia study

Table 4.4 Correlations between the Belief Clusters in Asia

	Belief Cluster 1	Belief Cluster 2	Belief Cluster 3	Belief Cluster 4
Belief cluster 1	1	0.2554	0.1785	0.3789
Belief cluster 2		1	**0.3390**	**0.4155**
Belief cluster 3			1	0.2914
Belief cluster 4				1

In interpreting the results, the evolution of policies in wildlife conservation in these countries was also taken into account. One limitation of the study is that the participant sub-sample from Japan clearly deviates from the research design. The Q-sample (Fig. 4.2) was administered, unfortunately, during the time of the Tohoku Earthquake in Japan. Partly due to this emergency, it was not possible, as originally planned, to interview participants from the Ministry of Trade and Industry (METI) and Ministry of Environment (MoE). As mentioned in the sections on the government-oriented and local NGO-dominated belief clusters, it was decided to keep the Japanese participants in the P-sample primarily because they were key participants in the WEMS system design.

The results of the analysis show low correlation between the factors or belief clusters (Table 4.4). Factor 1 (Belief Cluster 1) has a lower correlation with Belief Clusters 3 and 2; there is considerable overlap between factor 2 and the other factors 3 and 4, while factor 4 overlaps with factor 2 and moderately with factor 1. This implies that the factors represent fairly independent and authentically different views.

Grouping the ranking statements in these four factors gives a total explained variation of 65%. The factor ranking for each statement is provided in Table 4.5 and the individual's ranking represented in Table 4.6 corresponds to the averaged rankings of the four principal factors.

The four factors or belief systems can be named following the respective roles in their narrative structure as:

Table 4.5 Q Sort statements and their factor rankings in Asia

	Statement	Factor 1	Factor 2	Factor 3	Factor 4
1	Poachers should be shot when spotted within a protected forest area	2	−3	−2	−2
2	Wildlife should be protected as any kind of its use will lead to species extinction	0	2	0	−1
3	Wildlife is nature's gift which should not be substituted for economic subsistence	3	0	0	−1
4	Wildlife trade and hunting should be banned	3	1	−1	1
5	Conservation NGOs should be provided legal rights to stop wildlife crime	2	0	−3	−2
6	Conservation NGOs should be the custodian of wildlife crime database	1	−1	1	1
7	Wildlife policies should NOT allow options to cull endangered species when it causes crop and human damage	2	−2	1	1
8	Government prioritise wildlife conservation over wildlife trade	1	2	2	1
9	Local communities should be provided legal rights to hunt and sell wildlife for sustenance	−2	1	−3	−1
10	Wildlife policies should provide options to cull endangered species when it causes crop and human damage	−1	−1	−1	−1
11	Policies supporting sustainable use of wildlife are good to maintain a healthy wildlife population	−1	1	3	2
12	Protection of wildlife can only be in zoos and museums	−3	−3	−2	−3
13	Bush meat should be allowed as food for poor	−2	0	0	0
14	Wildlife is an economic commodity and should be utilised	−3	1	2	−3
15	Wildlife population can naturally revive and hence does not need significant restrictions on its consumption	−2	−2	−1	−2
16	Wildlife commodity should be listed in financial stock markets so it fetches uniform and better values	−1	−2	−2	−2
17	Science should determine the core policy on whether wildlife should be traded or not	0	−1	1	1
18	Scientific information should determine whether hunting should be allowed or not	0	1	1	2
19	Science should determine how nature should be managed	0	1	1	2
20	Wildlife cannot be conserved without addressing the rights of the forest community	1	2	0	2
21	Species listing of CITES appendices should be based on scientific information and not voting by member states	1	2	1	1

(continued)

Table 4.5 (continued)

	Statement	Factor 1	Factor 2	Factor 3	Factor 4
22	As long as utilisation of wildlife is regulated, wildlife remains protected	−1	0	2	0
23	Reputed academic institutions should manage wildlife crime databases	0	0	0	0
24	Only key wildlife species need to be protected as some can be traded	−2	−2	−1	−1
25	Marine species are losing out as protection of terrestrial mammals is getting precedence.	1	0	0	1
26	At certain point of CITES negotiation, trade-off of certain key marine species is considered better than allowing Tiger farms and ivory trade	0	0	1	0
27	Bird species are losing out as protection of terrestrial mammals is getting precedence	0	1	−1	0
28	Tuna should not be listed in CITES appendix as it is used as a major food product	−2	0	−1	0
29	Tiger conservation need more financial support than any other endangered species	1	−2	2	−2
30	Species-specific databases are most important than a general illegal trade database	0	−1	0	3
31	Tiger and Elephant crime databases will be enough for monitoring effective compliance of traded species	−1	−1	−1	0
32	Good funding for enforcement activities brings effective compliance of CITES	2	2	3	−2
33	The national wildlife crime information systems should be managed by a private IT company	0	−1	−2	−1
34	Professional IT consultants alone can bring out a good wildlife crime database	−1	−1	−1	−1
35	Governments should take the lead in addressing all issues with regard to enforcement and compliance of CITES	1	3	0	0
36	Governments alone are the custodian of illegal wildlife trade information	−1	−1	0	0
37	The UNEP-CITES secretariat should emphasise stricter sanctions against non-complying states	2	1	2	1
38	Harvest-related measures and trade-related measures should be used in tandem, to ensure the successful management of natural resources	−1	0	1	2
39	Conservation practices should also focus on income generation so as to achieve sustainability of such practices	1	3	−2	3

1. Strict conservation: ‘Bring in the local NGOs’.
2. Sustainable use: ‘Rely on bottom-up conservation by enforcement and compliance workers’.
3. State regulation: ‘Sustainable wildlife management means national regulatory policy’.
4. Expert-based wildlife management: ‘In science we trust’.

Table 4.6 Factor scores for each sort (ranking corresponds to the averaged rankings of the four principal factors) in the Asia Study (X represents highly loaded sorts within each factor; participants generating such sorts are called 'high-loaders', most 'typical' for a cluster/group)

Participant	Factor 1	Factor 2	Factor 3	Factor 4
#1 NGO (director) Japan	0.6712	0.3194	−0.105	0.0766
#2 NGO (president) Japan	0.8972X	0.0526	0.0245	0.0379
#3 NGO (volunteer) Japan	0.9151X	−0.0319	0.0622	0.1278
#4 retired government official (Japan)	0.5052	−0.1021	0.5047	0.0194
#5 CITES management authority (Thailand)	−0.0199	0.2359	0.7031X	0.2721
#6 CITES management authority (Thailand)	−0.1458	0.5157	0.5161	0.4484
#7 enforcement authority (law) (Thailand)	0.214	0.8346X	0.1116	0.1337
#8 Ministry of Natural Resources (Thailand)	0.1132	0.1195	0.8097X	0.0518
#9 Crocodile Farm Owner & Breeder (Thailand)	−0.3036	0.4823	0.2997	−0.0356
#10 international NGO (Thailand)	0.4805	0.2795	0.5780	0.0599
#11 international NGO (Thailand)	0.1138	0.7025X	0.3097	0.3404
#12 environmental lawyer (India)	0.4662	0.5434	0.0926	0.1685
#13 trustee (NGO) India	0.7444X	0.0376	0.1438	−0.2342
#14 CITES management authority (India)	0.0298	0.3073	0.0495	0.6618
#15 researcher MOEF (India)	0.6729	0.2680	0.3067	0.0507
#16 researcher MOEF (India)	0.6622	0.1229	0.4667	0.149
#17 researcher MOEF (India)	0.6047	0.2338	0.3306	0.24
#18 executive secretary (NGO) India	0.5932	0.3860	0.0753	0.4184
#19 wildlife journalist (India)	0.4088	0.1080	0.2315	0.7050X
#20 enforcement worker (India)	0.7355X	−0.0922	0.1265	0.3103
#21 enforcement worker (India)	0.6323X	−0.0799	−0.0695	0.1246
#22 technician wildlife mapping (India)	0.6439	0.4001	0.0331	0.4141
#23 Forest Community worker (India)	0.2635	0.7871X	−0.0035	0.2165
#24 NGO worker (India)	0.7157X	0.1201	0.0623	0.2413
#25 wildlife trader—India	−0.4962	0.5929	0.3601	0.1782
#26 scientist (India)	0.1581	0.1424	0.1296	0.7286X

4.5.1.1 Belief Cluster 1: Strict Conservation: 'Bring in the Local NGOs'

This belief cluster is embraced by Indian and Japanese local NGOs (see Table 4.6). It is the best covered factor with a population of 11 out of 26 participants. Excluded from sitting at the policy table due to their strongly ecocentric views, these NGOs are critical of government and rather sceptical about the role of science in wildlife policy. In an Asian setting, this is logical as science is frequently co-opted by governments; and science as a separate, independent sphere of life is a typical western idea, far less accepted, let alone practised in countries like India, Thailand and Japan (Danaher, 2008; Niraj, Krausman & Dayal, 2012; Zurcher, 2005). This group's ecocentric views, combined with anti-government sentiments, set its policy beliefs clearly apart from Belief Clusters 2 and 3, but with a moderate correlation with Belief Cluster 4 (see Table 4.4 on correlations between belief clusters).

In the policy narrative of this group of actors, *wildlife is nature's gift, so precious* as not to be degraded as a means for economic subsistence (statement #3, +3), even for feeding the poor. (Statement #13, −2, *interview data [P#3]): My basic understanding is wildlife is a member of ecological community of this earth. All species, including human beings, are members of the economic community of earth and hence utilisation of wildlife is against my idea.* The idea that wildlife could be treated as a usable and marketable commodity horrifies them (#14, −3); hence, they strongly *reject any form of sustainable use* (#11, −2), any conservation practice focussing on income generation (#39, −2) even for funding of conservation enforcement (#32, −1). Self-evidently, *wildlife trade and hunting should be completely banned* (#4, +3), and *harvest or trade-regulation-related policy measures are anathema* (#38, −1). Their ecocentric views strongly manifest themselves within their core beliefs where they consider *wildlife policies should not allow culling of endangered species even if they cause crop or human damage* (#7, +2), and that poachers caught in the act may be stopped even *by shooting at them* (#1, 2). Sceptical sentiments against national governments—reflecting the observation that government is not doing enough (in India), or may be exclusionary and captured by big business (in Japan)—are evident from the conviction that NGOs should be provided a role in enforcement matters. (Statement #5, +2, (*interview data [P#1]): If we talk about general conservation, I do not agree that government should take the lead in all enforcement matters. But if we take enforcement and compliance, the government alone can do the procedure. Collecting data and analysis can be done by NGOs.* Sceptical sentiments regarding the role of science among this group of policy actors are visible in the opinion, supported by this group only, that any national enforcement monitoring system should be managed by a non-biased entity—not by governments nor scientific institutions (#33, 0, all others reject this statement) *(Interview data [P #2]): The conflict within wildlife policy issues has risen to an extent that governments do not even listen to local NGOs. Government just needs organisations or scientists which support their views and not the reality.*

In terms of a policy narrative, it is clear why local NGOs are ascribed the role of heroes in protecting the victim (endangered species of wildlife flora and fauna). India experiences grave problems in enforcement of wildlife policies. The drivers of the problematic situation are, first, poaching by illegal traders and local communities banned from utilising wildlife in their own habitats; second, the conflict this has created between park managers (as representatives of the state) and local communities, and, third, between those who would allow use of the forest as a resource for the poor (utilisation) and those who give priority to protecting wildlife and declare its habitat as 'protected areas' (preservation). With so many 'villains', the local NGOs cast themselves as the only hope for effectively protecting endangered species; but this role as 'hero' in bringing about the moral of the policy story can only be played if they become more powerful policy players through legal rights, and with a more pronounced role in information collection and management (vis-à-vis science).

4.5.1.2 Belief Cluster 2: Sustainable Use: 'Rely on Bottom-up Conservation'

This group of policy actors is the second best covered in the P-sample, comprising 7 out of 26 participants. They mostly work in the 'down-to-earth' implementation levels of wildlife conservation in India and Thailand and include 'bona fide' wildlife traders/breeders, community workers and enforcement officials. As a group, people adhering to this belief cluster are clearly different from the *ecocentric* actors described in Belief Cluster 1. However, Belief Clusters 2, 3 and 4 show considerable overlap in the sense that they embrace anthropocentric values and beliefs.

The core beliefs of the sustainable users are that conservation practices *should not ignore wildlife as generator of income* (#39, +3), and that wildlife cannot be conserved without, simultaneously, *addressing the rights of local communities* (#20, +2). This aligns well with related anthropocentric views. They strongly resist poachers being shot as it is a violation of the rule of law. (*Interview data* [P#23]): *'I work with forest communities and fully understand that poaching has a social background. We are protecting animals and then we are shooting human beings?'* (#1, −3); they support culling endangered species that cause crop and human damage (#10, −2), and fearing the displacement of the local community, they strongly reject putting more money into tiger conservation (#29, −2); they also believe that the revival capacity of wildlife is strong enough not to necessitate consumption restrictions (on local communities) (#15, −2). Working on the compliance and implementation or output side of wildlife policymaking, they believe in the national government's responsibility to take the lead in wildlife management (#35, +3), and good funding as an indispensable condition for ascertaining CITES compliance (#32, +2). For this reason, they advocate that conservation should fund itself through sustainable use (*Wildlife is an economic commodity and should be utilised* [#14, +1]) and bring in protectionism at a time of unrestrained exploitation (#2 + 2). The traders believing in sustainable-use policy are concerned about wildlife losing its value if not traded (*Interview data* [P#9]): *'Wildlife trade should not be banned because people will then lose interest in animals and their value will be lost'*. This is similar to the views studied by Moore (2010) on elephant conservation: traders argue that conservation practices should also focus on income generation so as to achieve sustainability of such practices (#39).

Here, the heroes are the supporters of sustainable use with the motive of saving the local community and traders from *fortress conservation*. The victims in the story are the local traders and local community who depend on wildlife as a source of livelihood. The moral behind the story is that anthropocentric approaches alone bring income and livelihood for the poor local community, and this, in turn, will help conservation of forests and wildlife. For this reason, they reject participation of grass-root NGOs in information collection as they consider the information may be manipulated for ascertaining ecocentric beliefs. They agree with the CITES decision on barring NGOs in their involvement in WEMS (*Interview data [P#11]): Among NGOs, certain international NGOs alone are qualified in enforcement information collection and dissemination, but this should also be well monitored.*

4.5.1.3 Belief Cluster 3: State Regulation: 'Sustainable Wildlife Management is a Matter of National Regulatory Policy'

This belief cluster is comprised of 5 out of 26 in the P-sample (four from Thailand and one from Japan). Four are high-level government officials. The fifth represents an international NGO. As will become clearer later on, this group of policy actors differentiates itself from local NGO representatives, the implementation and compliance workers, and the scientific experts who, respectively, dominate the other three belief clusters. Their views clearly derive from official government policies on wildlife management, including being signatories to the CITES Convention. Their beliefs are also shaped by the traditional role of their governments in the overall political structure of power.

While the Indian government, operating in a fairly pluralist and democratic political system,[3] maintains a strong legal and policy framework to regulate and restrict wildlife trade, the line of semi-authoritarian (Chang, Chu, & Huang, 2006) Thailand is to maintain an open market policy towards wildlife trade matters. With the growing number of exotic animal breeders and safari parks, national agencies struggle to keep a balance between overexploitation of wildlife and its sustainable use. According to CITES National Legislation Project, Thailand, unlike India, has been rated as generally meeting the legislative requirements for the implementation of CITES and is considered in Category 1 (CITES, 2012a). However, some authors provide a different view because of booming illegal wildlife trade across the borders with neighbouring Myanmar, Laos and Cambodia, with several illegal wildlife markets within Thailand and just across the border (Nijman & Shepherd, 2011).

The Japanese government, operating in an open and democratic system, is nevertheless characterised as a 'managed democracy' (Estévez-Abe., 2003) where government and public policymaking is restricted to an 'iron triangle': a highly intertwined governance network of one political party in power (the Liberal Democratic Party, during most of the post-war period), big business and a strong bureaucracy. This 'iron triangle' of political, business and bureaucratic elites completely dominates public policymaking (Estévez-Abe., 2003; Muramatsu & Krauss, 1984).

[3] Freedom house ranking of India, Thailand and Japan

Country	Status	Freedom Score	Civil Liberties	Political rights
India	Free	2.5	3	2
Thailand	Partially Free	4.5	4	4
Japan	Free	1.5	2	1

(Scores are based on a scale of 0 to 7, with 0 representing Strongest and 7 representing Weakest performance.)

Freedom House (2012). 'Freedom in the World 2012: The Arab Uprisings and Their Global Repercussions'. Retrieved 17 March 2012, from http://www.freedomhouse.org/report/freedom-world/freedom-world-2012

The policy core belief of this government-oriented belief cluster is that *regulation* is key in wildlife protection (#22, +2), as without it protection of wildlife population would not naturally revive or survive (#15, −1). However, this is interpreted to mean a *right mix of preservation, conservation and trade*. They realistically state that wildlife is an economic commodity that should be utilised (#14, +2). *Sustainable use* of wildlife is their policy goal for maintaining a healthy wildlife population (#11, +2), using a well-balanced combination of harvest- and trade-related measures (#38, +1) that (at present) give priority to conservation over trade (#8, +2). *(Interview data [P#5]): 'In Thailand, a lot of people including traders believe in sustainable trade—why should we not think sustainably?'*

There is also a mild preference among them that regulatory effort should be directed centrally through the national governments. Local communities should not have hunting and selling rights (#9, −3); local NGOs should not have legal rights to stop wildlife crime (#5, −3, see also #20, 0) as upholding the rule of law is a traditional task of the government (#1, −2); and although governments are not the only custodians of illegal wildlife trade information (#36, −2), they clearly reject the idea that wildlife crime information could be outsourced to private sources (#33, −2, also #34, −1). *(Interview data [P#8]): 'NGOs do not understand why we are taking certain decisions. We have to work according to the law and not emotions. I am also not sure whether science can solve the existing conflicts in wildlife trade policy as science is seen invalid when it comes to political decision process—at least at an international level'.*

In order to keep their central role, governments are willing to subject themselves to stricter international CITES regulation (#37, +2) and, probably in this international framework, are strongly aware of budget constraints in wildlife management. They stress the necessity of good funding for enforcement and compliance (#32, +3). Although formally rejecting policies that focus on 'iconic' wildlife species alone (#24, −1, #30, 0, #31, −1), they clearly endorse more financial resources for tiger conservation (#29, +2).

This government-oriented policy narrative is fairly anthropocentric: wildlife is a natural resource to be exploited for human purpose. It is government's heroic task to guard these boundaries by ingenious management strategies that consist of expertly determined mixes of preservation, conservation and trade. Only in this way the policy story's moral—sufficient wildlife protection—can be brought about. This is not a task to be shared lightly with or outsourced to non-governmental players (as potential villains). Government's most urgent problem in wildlife management is funding for sufficient implementation, enforcement and compliance. On enforcement information-sharing, they strongly object to local civil society participation and having access to wildlife enforcement information.

4.5.1.4 Belief Cluster 4: Expert-Based Wildlife Management: 'In Science we Trust'

This belief cluster relies on the views of three scientists/information managers from India (see Table 4.6). All of them work outside government organisations. Yet, they operate at a national level, aiming to influence national and international policies on wildlife management. These actors base their beliefs on strong *rational* views[4] which emphasise that science is the basis for formulating pragmatic and optimal solutions for any policy problem. *(Interview data [P #14]): I entrusted former poachers who were part of the local community to be the custodian of forests. It worked and is now considered by FAO as a model in forest management... So Science-based solutions work'.* There is no principal preference for any solution which favours either wildlife protection or economic benefit. The solutions are deemed objective and value-free. Scientific institutions are capable of collecting sufficient and appropriate information to monitor and manage wildlife and wildlife trade. This is brought out clearly in the Q-analysis which identifies statements #2, #30 and #18 as the three statistically distinguishing statements for this belief cluster. The rationalist argument starts from the denial (#2, −1) that wildlife should be protected, as any kind of use will lead to species extinction. The protection of wildlife in danger of extinction is not an a priori fundamental concern, neither is the protection of specific species (#29, −2). Rationally, such preferences could only be validly stated if one had perfect information on all species. Hence, generation of species-specific databases is the key to scientific wildlife management; which, therefore, is more important than having a general illegal trade database (#30, 3). By having more and better information on specific species, scientific information could determine how nature should be managed (#19, +2) by (dis)allowing trade (#17, +1) or hunting (#18, +2).

In the policy narrative of the scientific group of actors, aggregated scientific information should be the basis for the wildlife trading policy (#30, +3). Such a policy should be the result of combining conservation benefits (#11, +2) with income generation opportunities (statement # 39, +3), including for the forest communities (#20, +2). There is also no specific a priori preference for cost-based policies (#14, −3), as the value of wildlife protection is not assessed in monetary terms (#16, −2). There is a strong belief that rational reasoning provides a more solid ground to convince governments on the relevance of CITES (# 19, +2) than market forces (#32, −1) or drastic compliance measures (#1, −2). This group sees a limited

[4] On the assumption that only humans can be rational. Other clear indicators of scientific experts' anthropocentric views are that, in their opinion, poachers should not be shot at the spot (#1, −2), and in wildlife conservation, the rights of the forest community should be taken into account (#20, +2).

role for NGOs in wildlife conservation (#5, −1) or in generating scientific information. However, they do not object to NGOs collecting enforcement information, though they do question the scientific rationality of an NGO-based information system.

The heroes in this policy narrative are the scientific experts and institutions. The villains are the governments which are currently incapable of rational decisions without the support of reputable academic institutions (#23, +1); they are ill-equipped and bureaucratic. The moral is that the policy-science interface needs significant improvement. With proper scientific information, the policies on wildlife protection will improve.

Looking at all statements pertaining to information and information management (i.e. statements ## 17, 18, 19, 21, 33 and 36), scientific experts cannot expect unwavering support from the other groups for their views. None of the groups shares the scientists' enthusiasm for giving priority in developing species-specific databases instead of a general illegal trade database (#30, 0, −1, 0, +3). Here is a clear example of a clash between scientific and practical research priorities. Local NGOs are outright sceptical about scientific expertise (esp. #17, 0; #19, 0). Implementation and compliance workers believe scientific expertise and information to be vital, but certainly not decisive (#17, −1), and governmental actors see a close alliance with science as instrumentally inevitable, but they are unwilling to hand over discretion and authority to science. They want science to be 'on tap, not on top'.[5]

Based on the above study, beliefs within the wildlife policy subsystem in India, Thailand and Japan can be summarised as follows:

(a) Belief Cluster 1 is ecocentric and represents the views of conservation NGOs.
(b) Belief Cluster 2 is a mix of neoliberals and pure anthropocentrics, representing the perspectives of the proponents of sustainable use.
(c) Belief Cluster 3 is authoritarian and represents the perspectives of government agencies.
(d) Belief Cluster 4 expresses scientific rationalism and represents the perspective of scientists.

It should be noted that in Asia, the science belief cluster (Belief Cluster 4) overlaps with both ecocentric (Belief Cluster 1) and sustainable use (Belief Cluster 2) clusters. The reasons for this overlap are described in more detail in Sect. 4.4.3 (on the comparative similarity and difference in the beliefs in Asia and Africa) and in Chap. 5 while showing the interaction of the coalitions.

[5] Mark Jones, director of a UK conservation NGO, says: 'The process of decision making has become intensely political. Parties choose to use scientific evidence to support their positions when it suits them, and refute the validity of the science when it doesn't.' (M. Jones, 2012)

4.5.2 Analysing Policy Beliefs in Wildlife Management in Africa

After analysing the beliefs in Asia, a similar study was carried out in East Africa, where the three countries (Kenya, Uganda and Tanzania) involved in the implementation of WEMS were considered as the study area. Due to limitation of funding, Congo-Brazzaville could not be included. The research methodology used in Asia was repeated in the countries with the exception that certain Q statements were 'slightly' modified to adapt to fit the specific East African wildlife policy context (see Table 4.3, #28, #29 and #31). The participants were identified based on their roles and responsibilities within the wildlife policy subsystem in the three countries of the study. The selected P-sample included 31 participants belonging to various stakeholder categories. The detailed configurations of the participants are provided in Fig. 4.3 and Table 4.9.

The Q-sort was administered with 39 statements to the participants in Kenya, Uganda and Tanzania. This was followed by an analysis using the PQ method. The details of the analysis are provided in the section.

An analysis of the Q-sort identified four belief clusters, or perspectives on wildlife management in East Africa. The study showed very low correlation between the beliefs. Belief Cluster 2 had the least correlation with the rest of the factors. There were considerable overlaps between Belief Cluster 4 and Belief Cluster 1 (Table 4.7).

The results of the factor analysis were presented by creating a descriptive label for each belief cluster (factor) accompanied by a narrative. In this instance, the narratives relate descriptions on various perspectives on wildlife management. The four perspectives are described in the following sections. Numbers in parentheses relate to the ranking of the statements listed in Table 4.8, and Table 4.9 provides the factor scores for each individual. While developing the narratives, the interview data were cross-checked by revisiting the interviews of people who loaded highly on each factor. Based on the Q-sort analysis and the interview data, the factor analysis results are described as follows:

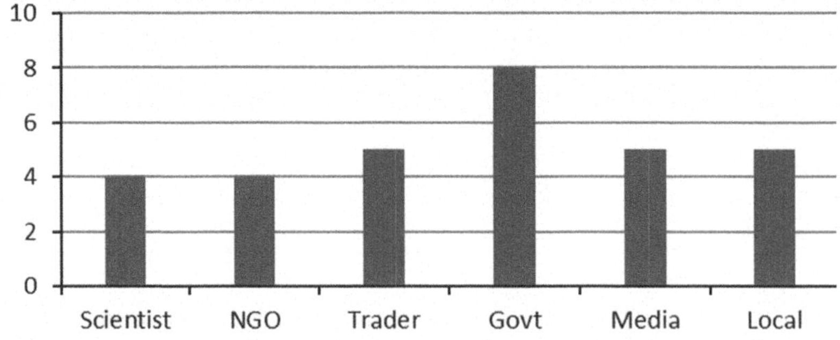

Fig. 4.3 Category of participants in the Africa study

Table 4.7 Correlations between the belief clusters (factors) in Africa

	Belief Cluster 1	Belief Cluster 2	Belief Cluster 3	Belief Cluster 4
Belief cluster 1	1	0.0889	0.2886	**0.4469**
Belief cluster 2		1	0.0002	0.2463
Belief cluster 3			1	**0.3552**
Belief cluster 4				1

1. Belief Cluster 1: Science Based: 'Science should take the lead'.
2. Belief Cluster 2: Ecocentric: 'Local NGOs should manage the commons'.
3. Belief Cluster 3: Sustainable traders: 'Sustainable trade—the only way forward'.
4. Belief Cluster 4: Hierarchists: 'We are the custodians and we decide'.

Table 4.7 shows very less correlation between the belief clusters, an indication of fewer consensuses between the perspectives of each coalition (<0.5).

Weighted averaging (Brown, 1996) of the scores was used to calculate each factor's (group's) average rating for each statement (Table 4.8).

To identify the individual participant's belief or perspective, each participant's ranking was transformed into a factor loading, signalling the degree to which an individual's ranking corresponds to the averaged rankings of the four principal factors (see Table 4.9).

Based on the above results, the characteristic feature of each belief cluster is defined as follows:

4.5.2.1 Belief Cluster 1: Science Based: 'Science Should Take the Lead'

This factor explains 18% of the variance in responses (Table 4.9). Among them, 6 participants are significantly associated with this factor (6/31). They include two traders, an NGO and a university professor from Tanzania, along with a Ugandan journalist and a Kenyan scientist. The most common feature of this belief cluster is their belief in science as an answer to maintain sustainability of wildlife. They strongly agree to policies supporting sustainable use (where science determines what can be traded or not) of wildlife as a good mechanism to maintain a healthy wildlife population (#11, +3); they also agree that policies related to wildlife trade should be dealt through science (#19, +2) and that scientific information should determine whether hunting should be allowed or not (#18, +2). This belief cluster reaffirms the characteristic feature of the perspective of the actors who consider scientific rationalism paves way to good decisions on hunting and wildlife trade. However, for some, the concept of science is utopian. As one participant (P #15, trader) asserted:

> When I say science it should be a perfect study. Science goes by proof. You start by theory, you put your hypothesis and you prove it beyond reasonable doubt, you have the facts and evidence. And if you come with this evidence, I (trader) cannot say anything about it. I am talking about science which is not manipulated. I am talking about an ideal situation. If you are interpreting of what I did not tell you, then it is not science.

Table 4.8 Q Sort statements and their factor rankings in Africa

	Statement Ranking	Factor 1	Factor 2	Factor 3	Factor 4
1	Poachers should be shot when spotted within a protected forest area	−1	−2	0	−3
2	Wildlife should be protected as any kind of its use will lead to species extinction	−1	3	0	1
3	Wildlife is nature's gift which should not be substituted for economic subsistence	0	1	−2	−2
4	Wildlife trade and hunting should be banned	−2	3	−3	−2
5	Conservation NGOs should be provided legal rights to stop wildlife crime	−1	1	0	−1
6	Conservation NGOs should be the custodian of wildlife crime database	−1	0	−1	−2
7	Wildlife policies should NOT allow options to cull endangered species when it causes crop and human damage	−2	1	0	2
8	Government prioritise wildlife conservation over wildlife trade	0	2	0	0
9	Local communities should be provided legal rights to hunt and sell wildlife for sustenance	1	−3	0	−1
10	Wildlife policies should provide options to cull endangered species when it causes crop and human damage	2	0	1	−2
11	Policies supporting sustainable use of wildlife is good to maintain a healthy wildlife population	3	−1	1	0
12	Protection of wildlife can only be in zoos and museums	−3	−2	−1	−3
13	Bushmeat should be allowed as food for poor	0	−3	−1	−1
14	Wildlife is an economic commodity and should be utilised	1	−1	3	2
15	Wildlife population can naturally revive and hence does not need significant restrictions on its consumption	−1	−1	−2	−1
16	Wildlife commodity should be listed in financial stock markets so it fetches uniform and better values	−3	−2	2	−1
17	Science should determine the core policy on whether wildlife should be traded or not	2	1	−1	3
18	Scientific information should determine whether hunting should be allowed or not	2	0	0	1
19	Science should determine how nature should be managed	1	2	1	1
20	Wildlife cannot be conserved without addressing the rights of the forest community	2	1	3	2
21	Species listing of CITES appendices should be based on scientific information and not voting by member states	1	2	−1	1

(continued)

Table 4.8 (continued)

	Statement Ranking	Factor 1	Factor 2	Factor 3	Factor 4
22	As long as utilisation of wildlife is regulated, wildlife remains protected	3	−2	2	1
23	Reputed academic institutions should manage wildlife crime databases	0	−1	−1	0
24	Only key wildlife species need to be protected as some can be traded	−2	−2	2	0
25	Marine species are losing out as protection of terrestrial mammals is getting precedence	1	0	−1	0
26	At certain point of CITES negotiation, trade-off of certain key marine species is considered better than allowing Tiger farms and ivory trade	0	0	1	0
27	Bird species are losing out as protection of terrestrial mammals is getting precedence	−1	1	−2	0
28	African fish species should not be listed in CITES appendix as it is used as a major food product	0	0	−2	−1
29	Elephant conservation needs more financial support than any other endangered species	0	0	1	−1
30	Species-specific databases are most important than a general illegal trade database	0	0	0	3
31	Elephant databases managed by CITES are the most trusted databases	−1	0	0	0
32	Good funding for enforcement activities brings effective compliance of CITES	0	2	1	1
33	The national wildlife crime information systems should be managed by a private IT company	−2	−1	−2	−2
34	Professional IT consultants alone can bring out a good wildlife crime database	−2	−1	−1	−1
35	Governments should take the lead in addressing all issues with regard to enforcement and compliance of CITES	1	2	2	0
36	Governments alone are the custodian of illegal wildlife trade information	1	−1	−3	1
37	The UNEP-CITES secretariat should emphasise stricter sanctions against non-complying states	−1	1	1	0
38	Harvest-related measures and trade-related measures should be used in tandem, to ensure the successful management of natural resources	1	−1	1	1
39	Conservation practices should also focus on income generation so as to achieve sustainability of such practices	2	1	2	1

Table 4.9 Factor scores for each sort (ranking corresponds to the averaged rankings of the four principal factors) in the Africa study

Participant	Factor 1	Factor 2	Factor 3	Factor 4
1 scientist, Kenya	0.7634X	0.0659	−0.0955	0.1287
2 government, Kenya	0.1078	0.7226X	−0.1642	−0.0958
3 education, Kenya	0.2055	0.1643	0.7352X	0.0935
4 journalist, nation media, Kenya	0.2782	−0.1879	0.0357	0.5623
5 government, Kenya	0.318	0.1055	0.7000X	0.2322
6 NGO, Kenya	−0.0672	0.8539X	0.0546	−0.0388
7 government (CITES officer), Kenya	0.4059	0.5625	0.1571	0.1558
8 journalist and communications director	0.4078	0.3152	0.1887	0.5867
9 trader, Kenya	0.0293	0.1775	0.4837	0.5508
10 (wildlife conservations society of Tanzania (NGO)	0.6337	−0.0139	0.3332	0.3567
11 scientist, Univ. of Darussalam	0.6192	0.2348	0.1575	0.3961
12 trader, Tanzania	0.6071	−0.164	0.5474	0.3239
13 government, Tanzania	0.5058	0.4715	0.303	0.3457
14 government, Tanzania	0.0688	0.0606	0.1402	0.7705X
15 trader, Tanzania	0.8088X	−0.0273	0.242	0.2833
16 journalist, Tanzania	0.2324	0.2095	0.5582	0.1654
17 scientist TAWIRI, Tanzania	0.2821	0.5103	0.1198	0.595
18 NGO, Tanzania	0.2735	0.0684	0.3713	0.5801
19 local community, Tanzania	0.2781	−0.462	−0.5014	−0.293
20 government, Uganda	0.3529	0.0926	0.3533	0.6506
21 trader, Uganda	0.3887	−0.2825	0.6706	0.0396
22 scientist, Uganda	0.2243	0.4331	0.1888	0.562
23 new vision, Uganda	0.6930X	0.2473	0.3489	0.2355
24 journalist, *daily monitor*, Uganda	0.4402	0.6073	0.3249	0.2625
25 local community, Uganda	0.071	0.6622	0.0244	0.3334
26 local community, Uganda	0.551	0.242	0.0966	0.3143
27 local community, Kenya	0.4783	0.4052	0.1155	0.4446
28 government, Uganda	0.2768	0.2965	0.1611	0.7494X
29 trader, Uganda	−0.152	−0.184	0.7704X	0.1971
30 NGO, Uganda	0.0881	0.7566X	−0.0309	0.2615
31 NGO, Uganda	0.66	0.5743	0.0739	−0.103
% expl. Var	**18**	**16**	**13**	**16**

(X represents highly loaded sorts within each factor; participants generating such sorts are the 'high-loaders', most 'typical' for a cluster/group)

Another participant (P#11) also agrees to science within the same terms: '*Science means perfect science…because in science also there is politics!*

This cluster defines a coalition of traders and scientists. Pro-trade governments and traders also consider an endorsement from science or scientists as crucial to deal with policies that are difficult to get accepted through conventional mechanism. For some, the term 'sustainable trade' is a backing of science-based trading process. According to one of the scientists (P#11): '*If you don't have scientific backing, how do you allocate quota (quota for hunting and trading)? The involvement of scientists with governments is an inclusive process. In some instances, there is not enough data for scientists to come to a conclusion but then somehow a decision has to be made. On the other hand, governments are also hesitant in including scientists in data collection and scientific research on wildlife trade. This is because, some governments (e.g. Tanzania) are interested in trading in wildlife, but science may provide a different picture. So, all these processes remain complicated*'.

Though participants of this belief cluster are sceptical on how governments use science or scientist, they still consider quantitative outputs as a tool for decision support. For instance, their reasons for disagreement to statements: '*Wildlife policies should NOT allow options to cull endangered species when it causes crop and human damage*' (#7, −2) and '*Wildlife trade and hunting should be banned*' (#4, −2), arise from the argument that numerical evidence should determine policies and not cultural or ethical values. This is clearly evident from one of the scientists' assertion: '*In some parts of Kenya, the problem of human deaths due to elephant attacks is also increasing. So it's like when the population builds up and when we close trade there is also a problem. When it comes to culling I may not be clear on that. There are some other ways. That's why I said science should be used to find out the numbers and then recommend the actions*'.

In spite of their belief in science, participants of this belief cluster admit that governments are the ultimate power groups which can overturn scientific reality. For this reason, they do not believe that stricter sanctions by international bodies like CITES can help bring this process under control (#37, −1). As one of the participants mentioned (P #15, trader):

> How can they (CITES) do this (on compliance)? They don't have any mechanism. How can they use sanctions without using governments? See, for example, if a person is found with rhino horn in Tanzania, he is arrested, but the same person can sell in Yemen. So how can stricter sanctions bring compliance?

From this policy narrative, science believers consider themselves as the heroes protecting human values through scientific truth.

4.5.2.2 Belief Cluster 2: Ecocentric—'Local NGOs Should Manage the Commons'

This factor explains 16% of the responses variance (Table 4.9). Four Q participants are significantly associated with this factor (4/31). They include two NGOs, one Kenyan Government official and a Ugandan journalist. The most common feature of this belief cluster is their belief in the intrinsic value of nature and wildlife. According to this belief cluster, wildlife trade and hunting should be banned (#4, +3), and wildlife should be protected, as any kind of its use will lead to species extinction (#2, +3). These core values determine the ecocentric approach of the belief cluster. Within the wildlife policy subsystem, they ask for conservation NGOs to be provided legal rights to stop wildlife crime (#5, +1), as they are the most reliable agents in information collection on poachers (P#2). Another participant (P #6) observes: *'Yes, they should be given legal rights to stop wildlife crime. We do not have enough sniffer dogs in this part of the world to check on consignments. So the only option will be to provide legal rights to NGOs'.*

In policy terms, they want government to prioritise wildlife conservation over wildlife trade (#8, +2) and that decisions should not be left to scientific facts alone but also to the cultural values of the society. Striking evidence to this belief is asserted by a Kenyan enforcement official (P#2):

> We should not leave to science alone to determine on whether trade should be opened or not. It should be complemented with the cultural value of the society.

They are also not in favour of providing legal rights to local communities to hunt and sell wildlife for sustenance (#9, −3), as they feel, if the rights are provided to the locals, then that will be the end of wildlife (P# 6). They totally reject consumptive utilisation of wildlife even when they are regulated (#22, −2); for them, all species should be protected and there should be no trade in them (#24, −2). They even do not agree to harvest-related measures and trade-related measures being used in tandem, to ensure the successful management of natural resources (#38, −1).

This belief cluster represents strict ecocentrics who opt for zero trade and zero consumption of wildlife.

The participants within this belief cluster indeed see a crucial role for NGOs as custodian of a wildlife crime database but do not deny the role for the government (#8, +2) in maintaining it.

4.5.2.3 Belief Cluster 3: Utilitarian—'Sustainable Utilisation—the Only Way Forward'

This belief cluster explains 13% of the response variance (Table 4.9) to which 4 participants are significantly associated (4/31). They include two Kenyan Government officials and two Ugandan Traders.

Unlike the other two beliefs, here they consider *wildlife as an economic commodity which should be utilised* (#14, +3). According to one participant (P#5):

> The basic policy on wildlife trade has not changed in East Africa since 1975 when wildlife was seen as a land use—something that should pay for its own (Kenya banned hunting and trading of wildlife in 1975).

Two other participants also echoed the same and according to a Ugandan Trader (P21), who loaded highly on this belief cluster,

> Ugandan laws will change and there will be soon wildlife ranches in here as we see in South Africa.

For all the above reasons, their views are strictly against the ecocentric belief cluster and strongly disagree on conservation NGOs being the custodian of wildlife crime database (#6, −1) or providing them legal rights to stop wildlife crime (#5,

−1). This is the only belief cluster which recommends wildlife commodity to be listed in financial stock markets so that it fetches uniform and better values (#16, +2). A trader (P#21) who loaded high on this belief cluster justified his support to the statement as follows:

> I agree to this so we can benefit if the prices are controlled. While we export an animal, we do not know the exact market cost abroad.

Similar views were expressed by (P#29) and (P#5) where they strongly emphasise that wildlife cannot be conserved without addressing the rights of the forest community (#20, +3). According to a trader:

> My forefathers never lived on livestock; they lived on consuming wildlife, so how can you address conservation without addressing their issues? (P#21).

> Though East African countries have policies that provide rights to local community, it is rarely practised (P#5).

They also consider themselves as the sole supporters of communities:

> When we capture these animals, and we provide incentive to community, they become friendly to animals (P#29).

According to this belief cluster, only key wildlife species need to be protected and some can be traded (#24, +2). They strongly oppose banning of wildlife trade and hunting (#4, −3) and governments being the custodian of illegal wildlife trade information (#36, −3). According to them, South Africa and Botswana are good examples of wildlife management where they promote trading and hunting (P#5, P#21 and P# 29).

An interesting aspect of this belief cluster is that they do not clearly oppose governments. But when it comes to information, they want the government to outsource some work to private entities:

> Government has all the facility, but it should offer 25-30% of their work to private Parties. For example, when you grow food, you have to leave some for the children and some for financial benefit! (P#29).

On all other terms, they are supportive of government policies, though some of the local traders feel marginalised by governments when it comes to wildlife trade.

Talking about science, they prefer political voting over science-based decision-making (#21, −1). There are two reasons for their support of politics: a) they do not consider scientific information to be true (P#29); and b) they consider that state sovereignty may be questioned if decisions are only science based.

> These infringe on the sovereignty of nations and you cannot surrender the sovereignty of a nation. We can use scientific information but the decision should be based on the best interest of the country. (P#5).

For this belief cluster, science is not perfect nor entirely reliable but is useful as a supportive tool for the decision-making process. So it should not be science alone, it should also include cultural and ethical perspectives.

In terms of policy narrative, this cluster clearly veers towards the anthropocentric perspective where human well-being should be considered prior to nature conservation. For this reason, they are strongly against any involvement of conservation NGOs in the wildlife management process. They do not prefer to rely purely on science as they think science at times undermines their objectives, and hence they prefer voting by member states on defining an outcome of policy. They also consider local culture as an instrument in supporting their arguments.

4.5.2.4 Belief Cluster 4: Hierarchists—'Government is the Custodian of the Wildlife on Their Territory'

This belief cluster explains 16% of the study variance (Table 4.9). Three participants are significantly associated with this belief cluster (3/31), which includes one Tanzanian and two Ugandan Government officials.

This group considers government as the single supreme authority to manage wildlife-related matters. They accept science as an instrument in the policymaking process (#17, +3), strictly to justify their arguments. As one participant noted:

> Science can determine what has to be done, like, for example, the size of the land and the population of species. So I agree to science-based decision process. (P#28).

They also don't agree to a comprehensive database as they think a 'species-specific' database would suffice the purpose of CITES compliance (#30, +3).

They also strongly oppose the fact that poachers should be shot when spotted within a protected forest area (#1, −3). Here, the reason is not anthropocentric alone, but more instrumental, where they fear losing information once a poacher is shot. According to (P#28): *'If you kill a poacher, you will lose information about his network. To plan better in future, we need to know the information about the weapon and his network. The main thing–we will lose out on information…'*.

This belief cluster also rejects involvement of NGOs and considers local NGOs as a threat to democratic processes. However, some support the views of 'certain' international NGOs and consider them as doing a good job (P#14).

On a policy narrative, the stakeholders within this cluster consider government as superior and righteous and as the only entity that can balance interests and manage all the actors within the wildlife policy subsystem. They do not object to science or scientist, but do not want to assign authority to them.

Based on the Q study, policy beliefs within the wildlife policy subsystem in Kenya, Uganda and Tanzania can be summarised as follows:

(a) Belief Cluster 1 expresses scientific rationalism.
(b) Belief Cluster 2 is ecocentrism.
(c) Belief Cluster 3 is anthropocentric, representing the perspectives of proponents of sustainable use.
(d) Belief Cluster 4 expresses authoritarian beliefs and represents a hierarchist perspective.

Similar to the Q study carried out in Asia, the 'hierarchist' belief cluster (Belief Cluster 4) overlaps with both scientific belief cluster and sustainable trade belief cluster. The reasons for this overlap are described in the next section.

4.5.3 Comparative Similarity and Differences in the Beliefs in Asia and Africa

An analysis of the results of the study in Asia and Africa reveals the following important features within the belief clusters:

Ecocentrism: Among the core beliefs, ecocentrism remains distinct in both Asia and Africa. Though the belief is primarily associated with conservation NGOs, it also doesn't mean that all NGOs are ecocentric. The belief cluster in Asia and Africa has clearly exhibited variation among NGO beliefs regarding conservation and trade. Ecocentric actors operating in the field of wildlife conservation (including different aspects of harm to animals) primarily develop their policies from an ideological point of view of and condemn activities that adversely affect wildlife. In order to resist such actions, they then seek venues to access higher profile audience in public policy for raising their concerns. Gaining attention from the public audience is a key, and for this reason, mass media and communication becomes an integral part of the campaign process. By doing so, they are also resisting trade decisions from being implemented. This causes isolation of the ecocentric belief cluster by the rest of the groups (a key finding which emerged out of the study). This distinction was quite clear in the studies carried out in Asia as well as Africa (see correlation between factors in Asia and Africa, in Tables 4.4 and 4.7).

The difference between the beliefs in Asia and Africa is that international NGOs in Africa are more resource rich and influential. This makes them a powerful resource base for influencing decisions within the wildlife policy subsystem. Earlier studies by Brockington and Scholfield (2010), as well as Garland (2008) confirm this case where transnational NGOs exert a certain power that can change the scenario. The influence of international NGOs within the policy process will be discussed later in Chap. 5. An important point to note here is that the prominence of ecocentric belief within a policy subsystem leads to disagreement among the rest of the belief clusters, especially on trading of wildlife or using it as a commodity. The two key statements used in the study clearly highlight the conflicts between the ecocentrics and the rest of the actors: *Policies supporting sustainable use of wildlife is good to maintain a healthy wildlife population;* and a more direct statement by CITES—*Harvest-related measures and trade-related measures should be used in tandem, to ensure the successful management of natural resources*[6] (which is one of

[6]This is a statement by John Scanlon, Secretary-General of CITES, during the 17th Special Meeting of the International Commission for the Conservation of Atlantic Tunas. http://cites.org/eng/news/sg/2010/20101119_sg_statement_ICCAT.php

the objectives of the CITES Convention itself). This is one area where the conflict between the ecocentric actors and the CITES convention becomes apparent. It should be noted that CITES does not advocate restriction on trade, whereas the ecocentric actors are against any form of wildlife trade.

Sustainable use: Another important finding from the study is the presence of hybrid groups within the wildlife policy subsystem. The empirical study clearly revealed overlapping policy beliefs among the hierarchists, the scientific rationalists and utilitarians (see Table 4.4 and Table 4.6); together, they contribute to a consensus belief cluster focusing on 'sustainability' of trade, use and conservation. One of the core characteristics of this belief is in their support for sustainable use and sustainable trade (common both in Asia and Africa). However, there remain some nuances which are different. For example, while the sustainable-use community commonly agree on the need for science in managing nature, some participants in Asia and Africa have different views on whether *species listing of CITES Appendices should be based on scientific information or on voting by member states.* The empirical study revealed that in Asia there is a strong support for science and data in the decision-making process which is also substantiated in debates at CITES meetings (decisions on ETIS and MIKE was supported by both Japan and China). In Africa, on the other hand, they do not believe that listing of species should be based on scientific information alone (Mark Jones, writing for BBC in 2012, clearly noted this point as well (M. Jones, 2012).[7] This shows that traders in Africa see an opportunity through voting to influence the trade decision process.

Since sustainable use and trade is a belief which is promoted by neoliberal NGOs, Governments, Scientists as well as Traders, it is not surprising that the actors within this belief cluster have different views on the use of science. It can be argued that, as Africa is a range state of several wildlife products and that it is one of the sources of national income, it is quite obvious that pro-trade groups would prefer less involvement of science. They are also of the opinion that harvest-related measures and trade-related measures should be used in tandem (support to CITES) and conservation practices should also focus on income generation so as to achieve sustainability of such practices.

In short, sustainable use is one belief cluster which is a mix of scientific rationalism and utilitarianism. For this reason, it is difficult to ontologically 'separate' or categorise the core belief. The concept of 'sustainable use' is advocated by governments, traders and scientists, all from their own standpoint, and this is common in both Asia and Africa. However, when we examine the international negotiation process, it is clear that the concept of sustainable use is more a neoliberal or utilitarian policy, where trade is at times justified through a process of *scientification.* This

[7] According to Mark Jones (M. Jones, 2012), '…the Zambian delegation rolled out Chieftainess Chiawa, head of a prominent indigenous group, to play the "poverty card" in support of their efforts to secure permission from the conference to down list their elephant population and sell off their stockpiled ivory; her pleas not to let her people starve when considering the fate of Zambia's valuable ivory stocks were impassioned, if somewhat lacking in logic.'

could be one reason why neoliberal NGOs find a place in the policy decision process on wildlife management while their ecocentric counterparts are not favoured. The role and prominence of this belief cluster will be clearer when we examine the interaction between the various coalitions within the policy subsystem in Chap. 5.

State Regulation: The third form of belief that emerged from the study was that of power, rule-bound or the hierarchist approach towards wildlife management. The study featured orderly rules of behaviour and authority structures among the hierarchists, created in order to avoid chaos, but with little faith in immanent self-organising or self-steering processes. This hierarchist belief is primarily (though not exclusively) associated with government representatives, where an authority defines how wildlife should be managed. Though the hierarchists emphasise on a rule-bound and a power-oriented structure, studies (Hood, 2000) indicate that the hierarchist belief can also at times work against them. According to Hood (2000), in many ways, top-level leadership is often quite weak in a hierarchist organisation where rules prevent every transaction from needing to be negotiated from a blank slate, for instance, in specifying what is to be worn or eaten in what circumstances, how people are to be addressed, how communications and meetings are to be handled. The rules also provide a basis for whom to blame when things go wrong—whoever didn't follow the rules (Hood, 2000).

Improvising Hood's argument and contextualising it with the hierarchist scenario within the case study, it can be said that, in Asia, the hierarchist approach is more about going by the rule and becoming inflexible in bending the rules to engage external actors on wildlife enforcement-related matters. According to them, information on wildlife crime should be managed by governments and enforcement agencies alone, and they are of the strong opinion that NGOs should be kept away from handling enforcement information (see Sect. 4.5.1). On the other hand, though the African Governments consider themselves as the most supreme authority on wildlife management, the rules are often flexed to engage external partners for funding needs. In the research it was clear that they strongly support the work of international NGOs and consider them to be doing a better job than their local counterparts. This demonstrates that hierarchism is strongly emphasised in Asia (at times authoritarian, as we see in Japan and Thailand) and severely weakened in Africa due to several economic and historical factors. Works of Garland (2008), Brockington and Scholfield (2010) clearly support these findings where they trace the reasons for weaker hierarchism to the roots of colonialism in Africa.

Scientific Rationalism: Another feature that reflects from the study is in the way various belief clusters interpret the use of science. In spite of the fact that all the beliefs have a faith in science, they do not fully agree on the way science functions. For this reason, the science belief has emerged as an independent cluster where very few participants in the study agreed to a positivistic science-based solution (see Fig. 4.4). Similar findings were observed by Andresen et al. (2005) while examining the role of precautionary principles in selected environmental regimes. According to them, scientists do have a legitimising role within environmental and resources regimes; however, their actual impacts are modest. Scientists are often listened to,

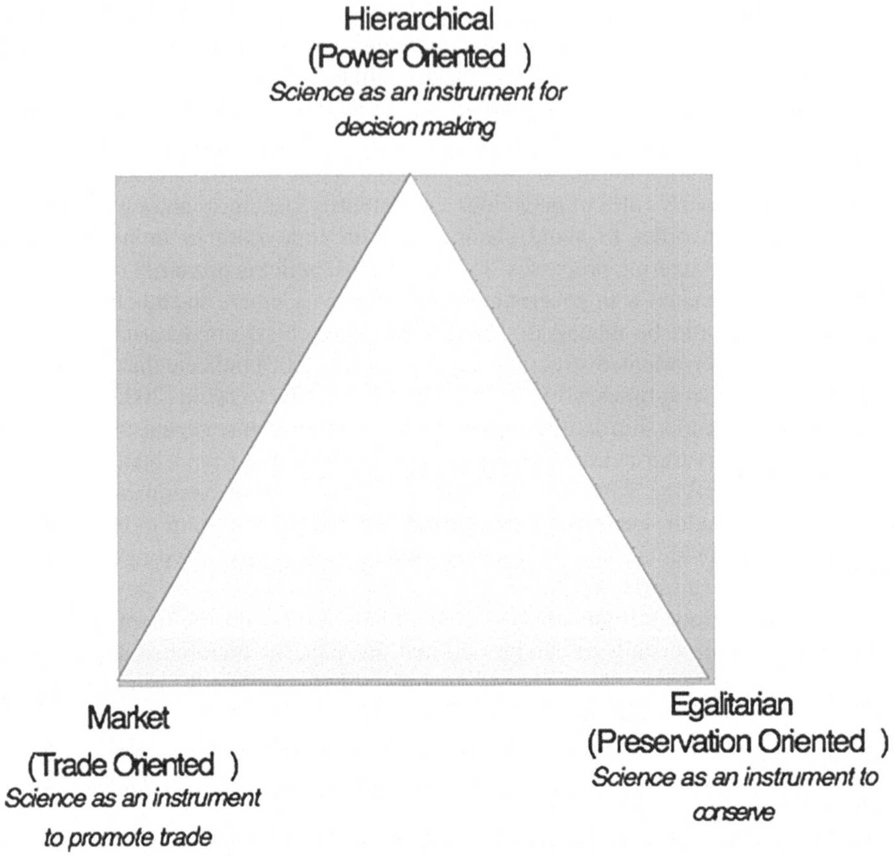

Fig. 4.4 The three beliefs within the wildlife-trade debate and their interpretation of science

but decisions are not entirely based on science instead more on other multiple decision criteria (Andresen et al. (2005)).

One should note that the science cluster is not a core belief and is more a representation of a secondary belief. It is also clear from the study that there is a strong similarity between the science beliefs in Asia and Africa.

As a summary, and as an explanation for the overlapping of belief clusters (see Table 4.4 and Table 4.7), it can be interpreted that ecocentrism is one core belief that can be easily distinguished from the study. The rest of the beliefs are hybrids of utilitarian values and science (sustainable use), as well as an association of hierarchical beliefs and science (State regulation). The scientific cluster is a representation of a secondary belief and is used by the ecocentrics, utilitarians and hierarchists to strengthen their position.

One of the advantages of using Q methodology is that one can isolate the various beliefs within the wildlife policy subsystem. It is also clear from the study that the ontologically separated beliefs are not part of any particular sector—NGOs, scientists, traders or government officials; rather, they are inherent constituents of core beliefs of actors within each sector. This clearly shows that professionals are not created solely based on their beliefs, but also based on their institutional rules or mission statements (cultural/structural) which then define their functioning. This applies to professional working in all the sectors, including NGOs. Hence, one can see divided positions even among NGOs addressing the same cause. The empirical study clearly outlines one important point—purely ecocentric (preservationist) approaches or purely utilitarian approaches will find it difficult to get involved in a policy decision process due to isolation by the other belief sectors. On the other hand, actors with hybrid beliefs can be more influential as policy interlocutors and play a significant role in the decision-making process. Other external factors affecting the policy system (for example, financial resources) then define which policy beliefs get prominence within the policy system.

4.6 Conclusion

In an attempt to address the research question posited in the beginning of this chapter, the empirical study in Asia and Africa analyses and maps out the various policy beliefs within the wildlife policy subsystem in Asia and Africa and explains their influence over the decision-making processes within the policy subsystem. The empirical case partially shows why certain core beliefs have difficulty in getting accepted within a policy subsystem and why there remain distinct gaps in official information-sharing between the stakeholders within the policy subsystem. This, in a way, outlines the reasons for the objection of WEMS-Asia by CITES when it involved ecocentric actors in enforcement information collection. Having said that, from an empirical point of view, it is also clear that the objection on WEMS cannot be exclusively attributed to one single cause, i.e. beliefs, and several interplaying external factors do have a role.

The study also indicates that *Science belief* has the potential in finding a common ground or congruence among several conflicting beliefs. It should however be noted that science does not change core beliefs. Which means, in many instances, actors (strongly positioned in their beliefs) choose the science that is convenient for them and reject that which is not, while coming up with scientific evidence favouring their position. In other words, though science is an important co-factor which is essential for a decision-making process (which could also be one reason for the adoption of WEMS in Africa), science alone cannot form a decision. It will need powerful political backing at every stage of its acceptance. But when an issue in context gets more polarised, scientists also get co-opted with powerful alliances for funding and personal benefits, which then leads to questions being asked about the validity of science in the decision-making process.

As a summary, it can be concluded that core and secondary beliefs do have a role or influence in decision outcomes. However, beliefs alone do not contribute to a particular decision. There are several factors external and internal to a policy chamber that influence coalition formation and thus the decision outcome. A more detailed explanation on how these advocacy coalitions function is provided in the next chapter.

Chapter 5
Explaining Policymaking on WEMS in Asia and Africa

5.1 Introduction

In this chapter, the research focuses on understanding the policy dynamics that led to both the rejection and eventual acceptance of the Wildlife Enforcement Monitoring System (WEMS) in different regions. Specifically, it examines the factors that led to CITES' objection to an NGO-supported WEMS in Asia and how, in contrast, the system supported by African governments was embraced in sub-Saharan Africa. The central question guiding this exploration is:

> What were the underlying policy factors that contributed to CITES' objection to WEMS in Asia, and what processes facilitated its subsequent adoption in Africa?

To address this question, the Advocacy Coalition Framework (ACF), as discussed in Chap. 3, serves as the theoretical foundation. The chapter demonstrates how competing coalitions of belief interacted within their respective policy subsystems and how these interactions shaped the divergent policy outcomes—the rejection of WEMS in Asia and its acceptance in Africa.

Before delving into the specific policy processes surrounding WEMS in both Asia and Africa, it is essential to understand the various types of policy subsystems, each defined by the behaviour of the coalitions within them. Weible and Sabatier (2009) identify three distinct types of subsystems:

Firstly, the *unitary subsystem*, where a unified and homogenous coalition exists, with little to no opposition, resulting in a decision-making process that is almost always unanimous. Secondly, the *collaborative subsystem*, where multiple coalitions hold differing views yet seek to reconcile these differences, necessitating cooperation between them. In such a subsystem, decision-making is generally based on consensus. Lastly, the *adversarial subsystem* is characterised by distrust and minimal cooperation between coalitions, where decisions are often hard to reach due to ongoing competition.

R. Chandran, *The Wildlife Enforcement Monitoring System*, Professional Practice in Governance and Public Organizations,
https://doi.org/10.1007/978-3-031-97084-9_5

It is important to note that, even within a collaborative subsystem, a shift in policy direction can occur when external factors undermine the core beliefs of the coalitions involved (Weible et al., 2010).

The objection to WEMS in Asia and its eventual acceptance in Africa can be understood through the lens of these policy subsystem types. Given that the two events unfolded at different times, the chapter addresses the core question in two distinct stages. Section 5.2 explores the reasons behind the rejection and redesign of WEMS in Asia, while Sect. 5.3 provides an analysis of the processes that led to its adoption in Africa.

5.2 Explanation of Objection and Redesign of WEMS in Asia (2005–2007)

The historical overview on CITES compliance information-sharing (Chap. 2) and the empirical case highlighted (Chap. 4) provide a clear picture of why certain non-governmental bodies are not preferred by certain CITES Parties. Those chapters also outline the implications this had for the Convention, where CITES Parties objected to civil society involvement in enforcement information collection and its subsequent sharing with the Secretariat, eventually leading to a lack of information-sharing on wildlife crime. The interviews with East Asian NGOs (ACA members) who were partners in the WEMS initiative also indicated similar approaches by their governments. For example, some CITES Parties were concerned about NGO participation in enforcement information collection and sharing because it might jeopardise their national and international policy agenda on trade related to CITES-listed species. This meant that there was a general scepticism about NGO involvement in enforcement data collection and which was well conveyed (detailed later in this chapter as well) by both CITES and national governments. To summarise, the objections from certain Parties and the CITES Secretariat to the NGOs involved in WEMS categorise the development of WEMS-Asia within an adversarial policy subsystem. In the particular case of WEMS-Asia, the coalitions were in conflict with each other where each entity was trying to influence decisions on the 'domestic-foreign' frontier.

In addition, and as mentioned earlier, CITES, the main policy broker in this game of negotiations, was going through serious resource shortages (see Chap. 2 and Chap. 3, Section 3.3.1), leaving them caught between making tough decisions but being well aware of the impact of these decisions on the Secretariat when it required Party contributions. How these external factors, outside of the policy subsystem, influenced the objection of WEMS by CITES is described in the next section through the lens of Advocacy Coalition Framework.

5.2.1 Advocacy Coalition Framework and Interactions within the Wildlife Policy Subsystem in Asia

The interactions within the wildlife policy subsystem in Asia in the context of WEMS development are highlighted using ACF, and as shown in Fig. 5.1. The figure presents an overview of the processes within the framework. On the left side are the two exogenous variables—one fairly stable and the other more dynamic. The stable factors influence the long-term coalition opportunity among the actors. And the dynamic factor influences the constraints and resources of the subsystem actors. The right side of the figure describes the actions within the policy subsystem.

5.2.1.1 Relatively Stable Parameters

From the empirical case described in Chap. 4 and the historical analysis of CITES enforcement provided in Chap. 2, it is clear that the discourses surrounding wildlife and forests in Asia and Africa are related to a struggle to classify its use,[1] management and ownership.

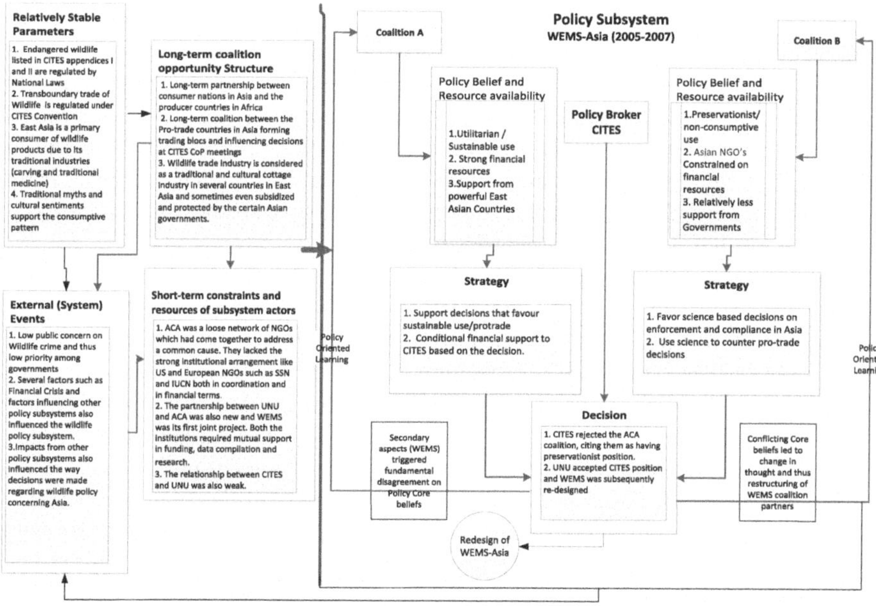

Fig. 5.1 Advocacy Coalition Framework and interactions within the wildlife policy subsystem in Asia

[1]The use of wildlife and forests is defined based on non-consumptive use, consumptive use and sustainable use. Non-consumptive use refers to preserving wildlife in their natural state, and income is generated either from watching, feeding or photographing them. Consumptive use refers

Within the context of the case study and in ACF terms, the *basic attribute* of biodiversity in Asia can be considered as vulnerable (Sodhi et al., 2004, 2009), which has then led to a stricter control in its management, thereby bringing them under the direct regulation of the state. The *basic distribution of natural resources* is also declining in Asia with consumers seeking resources in Africa for meeting their demands. The extraction and trade of certain wildlife resources such as elephant ivory, rhino horn and tiger products, that are part of *fundamental cultural values* and traditional practices in some Asian countries, are also regulated by national laws and internationally through the framework of CITES Convention. This stricter regulation has then led to flourishing of illegal trade. As more countries joined CITES, the struggle to impose limitation on the producer-consumer trading mechanism often turned CITES COP's into a battleground for winning positions on preservation- and utilisation-oriented policies. As per CITES regulation, for wildlife trade to take place between producer and consumer countries, both the parties should undertake appropriate compliance measures so that trade does not undermine the existence of the species. In other words, for trade to be approved by CITES, national wildlife law enforcement and compliance efforts should be adequate in both importing and exporting countries. It is often difficult to meet the complete requirement of CITES compliance process as the *basic constitutional structure* on the regulation of trade and conservation is often in the hands of an environmental or trade agency, whereas the role of on-ground enforcement is usually the responsibility of an enforcement agency. But while proposing trade, governments often report of complete compliance to the convention, ignoring non-compliance to the convention. NGOs, on the other hand, flag these issues prompting tensions and conflicts between the governments and NGOs. For example, NGO reports on the laxity of domestic legislation on internal ivory trade in Japan and China have forced CITES to carry out verification missions in these countries to investigate whether the internal control mechanism in these countries is adequate.

Within the circumstance of WEMS development, it can be concluded that for decades the policy subsystem in Japan and China has been more 'pro-trade'. A primary reason for this trend has been the close-knit relationship between business and bureaucracy in these countries in protecting traditional industries which depended on wildlife products. The empirical studies conducted in Asia also confirmed this pattern and explain why bureaucrats from China and Japan seek to promote wildlife trade during international conventions. The role of civil society in China (Ho, 2001) and Japan (Schwartz, 2002) is considered weak and environmental NGOs, especially in China, have the space to manoeuvre as long as they do not interfere with national policies (Ho, 2001). Schwartz and Pharr (2003), though confirm the historic neglect of civil society in Japan, argue that there is a new wave of modernisation taking place in Japan which is slowly changing the way NGOs are considered by the society. However, the empirical study carried out in Japan (see sect.

to using the wildlife for economic gain. Sustainable use is about creating a balance between the utilization and preservation of wildlife.

4.5.1—*Belief Cluster 1*) revealed that the conflicts within wildlife policy issues have risen to such an extent that governments do not listen to local NGOs.

5.2.1.2 External [System]Events

Changes in Socio-Economic Conditions and Technology

From the previous section, it is clear why the policy subsystem was not conducive to the needs of the ACA-UNU coalition. Apart from the non-conducive environment for an NGO-based coalition, there were other factors which were influencing the development of WEMS. Among them, the one that had a direct influence on WEMS development was the socio-economic situation within the wildlife-trade policy subsystem. WEMS was developed during the time of the financial crisis and when governments, businesses and civil society, all in parallel, were looking for funding resources to meet their own functional requirements. Shortage of funds also affected the functioning of UN agencies including UNU and CITES Secretariat. By 2005, Japan reduced its contribution to UNU and CITES quoting its weak fiscal condition. This meant that there were fewer resources within the subsystem and the respective coalitions were seeking external funds to carry out their activities. ACA, UNU and CITES all depended on external funding to operationalise their activities.

Though the financial crisis was a major impeding factor in the development of WEMS, it was fortunate to receive in-kind grants (GIS software) from ESRI corporation and monetary aid from International Fund for Animal Welfare (IFAW). These grants helped to create the WEMS prototype. However, information system designers barely looked into the political events that influenced the scaling up of WEMS-Asia.

Changes in Public Opinion

A key issue which influenced government decisions in Asia and a reason for them to move away from prioritising wildlife crime was the lack of public concern and the prioritisation of mainstream crime by bureaucratic elites (see also Elliott, 2007; Zhang et al., 2008). Though WEMS was developed as a tool for monitoring wildlife crime in Asia, the response to such a tool from governments was also weak.

Changes in Systemic Governing Coalitions

In Japan, for most of the year, the Liberal Democratic Party has been in power and supportive to pro-trade policies. The Chinese government has also been a strong supporter of wildlife trade. In spite of the change in political factions in Japan, the policy of the government towards wildlife issues has remained consistent for decades. For instance, Japan continued its cordial partnership with China during the CITES meetings, though the two countries disagree on several other trade-related issues discussed within the World Trade Organisation.[2] A shift from the Bern Criteria in 1976 to the Fort Lauderdale Criteria, adopted in 1994, clearly defines the role and strength of the pro-trade Asian coalition in bringing out this change from a preservationist approach at CITES to a more utilitarian one.

Policy Decisions and Impacts from Other Policy Subsystems

In addition to the earlier mentioned constraints, impacts from other policy subsystems also influenced the way decisions were made regarding trade and conservation of wildlife. For example, in many countries, agricultural and fisheries policies interlink with environmental policies. Lobbyists within the agricultural and fisheries sectors also play a significant role in influencing decisions at CITES COP meetings. Sometimes, a decision by a CITES Party to support the proposal of another during a COP meeting can be for maintaining diplomatic relations in order to foster a larger goal within another policy subsystem.[3] Such exogenous factors did have an impact (directly or indirectly) on the coalition formation and thus on the policy development of WEMS, as can be seen when the interactions within the policy subsystem are explained.

5.2.1.3 Long-Term Coalition Opportunity Structure

It was mentioned earlier (sect. 5.2.1.1; *Relative stable parameters*) about the pro-wildlife trade policies of China and Japan and its dependence on the resources from other regions including Africa. This common approach has created a long-term partnership between China and Japan in promoting wildlife trade globally. In addition, the traders (primarily trading associations) in these countries, holding strong political influence, acted as channels to their counterparts in several wildlife *Range States* and wildlife exporting nations. The traders (and their associations) also had substantial influence on their national governments in bringing out decisions favouring trade. This has in fact helped create a long-term coalition opportunity between the two countries thereby building a powerful trading bloc in Asia, influencing decisions at CITES COP meetings by lobbying wildlife range states and forming alliances with similar pro-trade nations in other regions.

---

[2] A WTO report indicates the conflict between China and Japan within the WTO agreement.
China—Measures Imposing Anti-Dumping Duties on High-Performance Stainless Steel Seamless Tubes ("HP-SSST") from Japan https://www.wto.org/english/tratop_e/dispu_e/cases_e/ds454_e.htm

[3] The breakout of mad cow disease in 2003 had caused critical problems for the US beef industry, and US diplomats in Tokyo were in intense negotiations with the Japanese Government to lift the ban on US beef. For this a softer stand on bilateral relations was important to restore the beef market (see Tōgō, 2010; Hanson, 2004). It is difficult to judge the influence of the diplomatic impasse on 'mad cow' disease and its relation with decisions on global wildlife trade. However, considering the timing, it should be noted that both UK and US governments did not object to the one-off trade of ivory into Japan and China despite strong opposition from UK- and US-based conservation NGOs.

5.2.1.4 Short-Term Constraints and Resources of Subsystem Actors

While the coalition of pro-trade factions in Asia was stronger, the coalition between conservation NGOs in Asia remained weak. ACA was a loose network of NGOs, which had come together to address a common cause. They lacked the strong institutional arrangement which western NGO networks such as SSN and IUCN have, and were also weak in both coordination and financial terms. The partnership between UNU and ACA was also new and WEMS was its first joint project. Both the institutions required mutual support in funding, data compilation and research. Hence, both UNU and ACA had to address these constraints through a joint partnership.

The relationship between CITES and UNU was also weak. UNU had never previously worked on a CITES enforcement-related project and its communication with the Secretariat remained at a minimal level. There was also no consultation between UNU and CITES concerning the development of WEMS.

5.2.2 Policy Subsystem

5.2.2.1 Coalitions, Policy Beliefs and Resource Availability

One point that is clear from the earlier sections is that, in 2005, the year when WEMS was being developed, there were several factors influencing the policy subsystem. In addition, the elephant ivory issue did have a significant influence on the decisions on enforcement monitoring systems. With the ban on ivory trade, there was a move towards approving a 'one-off sale' of ivory from African countries which had shown relatively good national law enforcement on controlling illegal trade of ivory. The export of the ivory was supposed to be designated to China and Japan. The procedure towards approving such a trade started in 2002 (see Chap. 2). One of the preconditions for approving the trade was a verification exercise in China and Japan by the CITES Secretariat to check whether their internal laws on enforcement and compliance met the prescribed standards of the Secretariat (CITES, 2005). This meant that both China and Japan had to demonstrate that they had taken internal measures and counter-measures to control the illegal trade of ivory. Understanding that ivory sale will be on the agenda for CITES COP 14, conservation NGOs were protesting against the trade of ivory into Japan and China, cautioning that a one-off trade would trigger widespread illegal trade.

The funding allocated for CITES activities was also considerably reduced due to economic meltdown in the USA, EU and UK. This weakened the CITES Secretariat's position in robustly standing up against Party non-compliance, as it also had to seek the support of member states in order to increase its operational budget and funding for its trust fund.

The scientific experts from UNU, on the other hand, allied with ACA where they worked as a single coalition in developing WEMS as a transparency tool to address the decision-making process of CITES. One point that should be noted here is: though ACA NGOs belonged to the preservationist camp, the information they collected were from government agencies. There was also a verification process defined within the information collection process where credible evidence alone goes into the system. Which means, irrespective of their position, the intention of ACA members was not to disrupt a policy process but it was more about bringing out grass-root information to the policy table as they considered Asian governments are not opening up the information on illegal wildlife trade. This move, especially within the context of ivory trade, went against the interest of the UNU-NGO coalition.

Strategy

As mentioned earlier, the development of WEMS was caught up between the needs of two coalitions:

(i) Pro-trade nations in East Asia who were particularly concerned about the effect of a ban on certain wildlife products (for example, ivory) and its subsequent impact on their traditional industries.
(ii) NGO coalitions who were concerned about the rise in illegal trade and thereby wanted to bring out the reality of the extent of illegal trade, especially in key species such as elephant and rhino.

On the other hand, a shortage of financial and staff resources at the CITES Secretariat (Policy Broker) meant that it needed to gain the support of its Parties (more than the NGOs) for proposing a nominal increase in its budget (CITES, 2007d). The strategy they used to influence decisions is outlined in the next section (see also Fig. 5.1).

Decision

Gupta (2010) empirically clarifies on how, in spite of the recurrent ideology of promoting transparency in the global environmental domain, the embrace of transparency is often used to promote or serve the often-conflicting interests of multiple actors. CITES's fear of the UNU–ACA coalition can be explained in a similar way where it was concerned that ecocentric actors may use the transparency channel to influence policy decisions at an international level. By objecting to WEMS, CITES effectively denied them the rhetorical materials that could have been used to craft a rebuttal to CITES decisions. A discussion during the fifty-fifth Standing Committee meeting (CITES, 2007c) clearly outlines this concern:

The Secretariat introduced document SC55 Doc. 10.1 (Rev. 1), reporting on the progress with assessing Japan's ivory trade controls and pointing out that some information provided by non-governmental organisations (NGOs) regarding Japan's ivory trade controls was misleading and inaccurate. The Secretariat stated that Japan's situation was satisfactory, as it had been at the time of SC54, and noted the impracticability of a database or a system that would allow to trace ivory products back to an original tusk in range States, which some Parties and NGOs asserted as necessary.

The objection of WEMS by the CITES Secretariat reinforces how MEA Secretariats can influence decision-making processes on global environmental governance. Although UN regulations define MEA Secretariats as neutral actors, they, at times, have to deviate from their neutral interlocutor role towards being an intermediary which favours one Party over the other. Studies by Sabatier (1988) also support this particular characteristic feature of national and international policy brokers. A more recent work by Jinnah (2014) further confirms this 'policy bend' by CITES in her book, *Post-treaty Politics,* where she explains the influence of the CITES Secretariat due to the 'historical privilege' that it had on making recommendations on proposals of trade, conservation and enforcement. Jinnah (2014) further explains that CITES influences mainly through knowledge brokering and negotiation facilitation within CITES politics. The influence of CITES as a policy broker becomes clearly evident when an issue in context becomes highly political or polarised as can be seen in the case of ivory trade or enforcement information-sharing.

From the above analysis, it can be concluded that the objection by CITES to WEMS-Asia could have been prompted by a fear of preservationist NGOs holding on to (or empowering themselves with) a set of information which could create an imbalance in the process of decision-making during CITES COP meetings (for example, the ivory decision which CITES had already approved). Stone (2002) expounds this phenomenon as a character of 'Polis', where information is interpretive, incomplete and strategically (at times, deliberately) withheld, and interpretations are more powerful than facts. Comparing this with the recent studies on the impact of crowd-sourced information (Shirky, 2011),[4] one can infer that the reasoning behind the objection by CITES could be due to fears of information bias by NGOs that could pose immense challenges to a planned or structured process of policymaking.

Policy-Oriented Learning

In the wake of the objection, UNU disassociated themselves from ACA and sought a new process of data-input mechanism involving government agencies. There are two reasons for this shift. Firstly, for WEMS to move forward, the NGO coalition may need to adhere to the needs of their governments. However, ecocentric NGOs would not prefer to change their core position or mission statement (core belief) especially on critical wildlife trade issues. Secondly, the shift in approach by UNU proves that scientific values are not based solely on ideology but also seek collaborative forums in order to progress their science forward. The shift by scientists can be considered as a process of policy-oriented learning, where they realised that it is almost impossible to have an enforcement monitoring system involving NGOs as information-gathering agents.

[4]*According to Shirky, 'Social media have become coordinating tools for nearly all of the world's political movements, just as most of the world's authoritarian governments (and, alarmingly, an increasing number of democratic ones) are trying to limit access to it'.* (Extracted from the article by Clay Shirky, 'The Political Power of Social Media—Technology, the Public Sphere, and Political Change'; published by *Foreign Affairs* magazine, Jan/Feb 2011 issue.)

5.2.3 Analysis of the Objection and Redesign of WEMS-Asia

From the objection and redesign of WEMS-Asia, one can conclude that the use of data or information within the wildlife policy subsystem has been more concerned with convenience than with finding a solution. Rather than identifying what the information tells them, respective coalitions are more concerned with how this information will affect or define their power or position. This particular view has been the main reason for the failure of WEMS when it worked in partnership with civil society networks. A second reason for the objection can be attributed to funding—a major external factor which continues to influence decisions within the wildlife policy subsystem. As mentioned in Chap. 4, the 2006–7 financial crisis, and the subsequent complications brought about by the shifts in global financial markets, led to challenges for major conservation efforts. The fact that the CITES Secretariat was proposing a nominal increase in its budget (CITES, 2007c), meant co-opting certain needs (if not all) of major donor countries, including China and Japan. It is therefore quite understandable that data on enforcement matters from an ecocentric group of actors, legitimised through a UN body, were the last thing that CITES would have wanted. Having said that, the extent of the influence exerted by financial instability within CITES and its relationship to CITES decisions remains difficult to judge and would require a separate investigation of its own.

The role of science and rationality in decision-making is worth noting. From the empirical study carried out, it appears that the role of science and rationality becomes a priori for decision-making only during a time of uncertainty and at a time when information is lacking (see also Chap. 4, Section 4.5.3—*scientific rationalism*). Even during times of uncertainty in decision-making, science cuts across all the various belief systems, with each coalition using scientific information as an authority to substantiate or gain support for their respective policies (see Chap. 4, Fig. 4.4). Clear examples of such an approach can be seen in the interactions during the complex decision-making processes on elephant ivory, whaling and wildlife crime. In these situations, each set of coalitions engages with reputed scientific institutions, formulating their own means of information collection, analysis and communication. Major news media, including *The Economist*, *Reuters*, *The Guardian* and *The New York Times,* were all used as science-communication channels of respective positions,[5] particularly at a time when clear scientific evidence was missing.

A lesson that can be learnt from the WEMS-Asia experience is: scientific information alone cannot bring about a shift in policy approach when the policy

[5] Examples of NGO positions conveyed through the media:

1. Trade in Wildlife—'Just let them get on with it', based on the report, 'Trading nature', by Dilys Roe, TRAFFIC, 2012; http://www.economist.com/node/11458275 (Access date March 2014)
2. 'Conservation-Trade Protection', *The Economist*, 5 March 2013; http://www.economist.com/blogs/babbage/2013/03/conservation (Access date March 2014)
3. 'The Price of Ivory', *The New York Times*, 31 December 2012; http://www.nytimes.com/interactive/world/africa/the-price-of-ivory.html (Access date March 2014)

subsystem itself remains highly polarised. A process of depolarisation can only happen when the issue in question (illegal trade) affects both the coalitions. Even though coalitions will still maintain certain positions that they never want to compromise (core beliefs are difficult to change; see Sabatier, 1988), a solution can emerge when coalitions identify a narrow common ground which provides them with scope for consensus and collaboration. In other words, for WEMS to be adopted in Asia, there should be a certain level of depolarisation within the policy subsystem to create room to identify common ground or—in policy terms—a boundary arrangement that would transform an adversarial policy subsystem into a more collaborative one (see Chap. 6).

In the next section, the adoption of WEMS in sub-Saharan Africa is outlined.

5.3 The Collaborative Subsystem in Sub-Saharan-Africa and the Adoption of WEMS (2010–2014)

In contrast to the situation with WEMS-Asia, the adoption of WEMS in sub-Saharan-Africa came with a welcoming gesture. To understand the reasons on why Lusaka Agreement Task Force (LATF) sought the support of an extended international research community, it is important to analyse the exogenous factors that influenced the decision of the Lusaka Agreement Governing Council and, the interactions within the policy subsystem that later led to the acceptance of WEMS in sub-Saharan Africa. These interactions are explained in the next section using ACF as a theoretical framework.

5.3.1 Advocacy Coalition Framework and Interactions within the Wildlife Policy Subsystem in Africa

The interactions within the wildlife policy subsystem in sub-Saharan Africa, particularly in the context of WEMS, are highlighted in Fig. 5.2. The description of the parameters of the figure is similar to that described for WEMS-Asia in Fig. 5.1.

5.3.1.1 Relatively Stable Parameters

The *basic attribute* of African biodiversity, though vulnerable, is better when compared with Asia in terms of genetic variation within populations; the number, relative abundance and uniqueness of species; and the variety, extent and condition of ecosystems. However, the policy problems influencing the management of wildlife and forests in Africa are similar to the ones in Asia where there remains a struggle to classify *exclusion* and *subtractability* of natural resources. The Asian and African

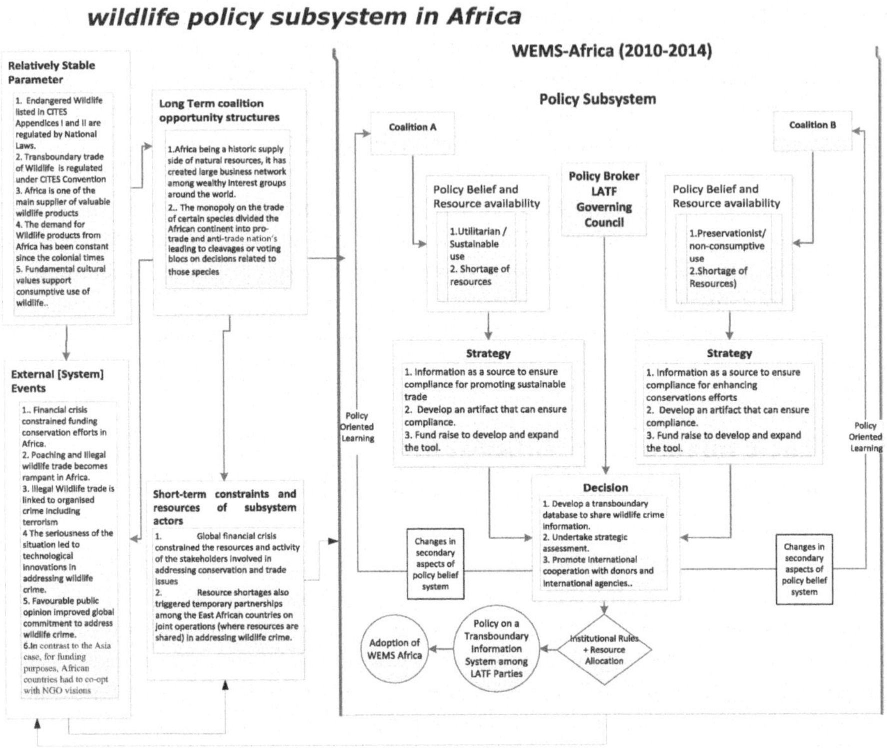

Fig. 5.2 Advocacy Coalition Framework and interactions within the wildlife policy subsystem in Africa

wildlife policy subsystem does get 'nested' whenever the demand for wildlife products from Asia influences its supply from Africa. One challenge in having a common decision strategy regarding a transboundary African wildlife policy has been related to the differences in the *distribution* and management of wildlife resources in different African countries. For example, while the African elephant population in the Southern African countries is well managed, the East, West and Central African countries are facing serious problems related to poaching, deforestation and habitat loss. The *constitutional structure,* as well as the rules and regulations, also varies from country to country, and this has often led the CITES Convention to favour African countries that have better governance in managing trade of wildlife and to impose restrictions on countries whose resources are ill managed.

Wildlife conservation policies are also influenced by the *fundamental cultural values* in Africa. For example, among certain native communities, wildlife is a food or an income resource which then questions preservationist approaches to conservation. The Communal Areas Management Programme for Indigenous Resources (CAMPFIRE) programme initiated in Zimbabwe, for instance, seeks to empower rural communities for conservation and sustainable development through harvesting of natural resources (Child, 1996). The majority of CAMPFIRE profits come from leasing trophy hunting concessions to foreign hunters. In the same vein, decolonisation and the demand for wildlife products from wealthy East Asian markets (see Zhang et al., 2008) are another factor that has to be taken into account. This marks a shift in the approach in wildlife trade from a 'European' and 'American'-centric approach to one that's more Asia-centric.

All these issues have made conservation and trade of wildlife in Africa highly contestable. The long-standing debate on African elephant conservation is a reflection of this complexity.

5.3.1.2 External (System) Events

Changes in Socio-Economic Conditions and Technology

Contextualising wildlife policy making in sub-Saharan Africa during the time of development of WEMS, it can be observed that several dynamic events (see Fig. 5.2) influenced decisions on wildlife conservation and trade in Africa. These events are described as follows:

(a) Financial Crisis: The 2007 financial Crisis, in general, had a major impact on conservation efforts all over Africa. Though wildlife remains as a major economic resource for sub-Saharan African countries, central governments hardly allocate significant funds for wildlife management, given the many other competing demands for governmental resources (see Newmark and Hough (2000)). Recognising this problem, and as a practice, several of the wildlife agencies functioning under the direct portfolio of the central government, including Kenya wildlife service, Uganda Wildlife Authority, etc., resort to external grants and aid for carrying out some of their activities. A point to be noted here is that most of the foreign-aid provided to Africa is not for wildlife conservation efforts but more for development-related work. With the 2007 financial crisis, the flow of aid resources from the USA and Europe into sub-Saharan Africa decreased and so did the income for conservation efforts. Several of the sub-Saharan African governments resorted to Chinese-Aid for development assistance which marked a shift from 'Western' dependence to a more 'Eastern' support. As the aid from China was primarily for infrastructural development (see Dreher and Fuchs (2015)), meeting conservation demand required wildlife agencies to think of new ways of managing their resources. The path they took in addressing this crisis is explained later in sect. 5.3.2.

(b) Illegal Wildlife Trade: The shortage of resources for conservation efforts further led to poor management of national parks, especially within the sub-Saharan region. Poaching and illegal trade of elephant and rhinos for their tusk and horns increased to an extent that it was threatening the survival of the species. In the wake of the seriousness of the issue, global bodies called for African countries to put their Parks in order.[6] For avoiding to be sanctioned, sub-Saharan African countries needed to demonstrate to the international community that their ground enforcement activities are in place. The news of the rampant illegal trade spread fast through social media where several NGO and media investigative reports exposed the level poaching in Africa. Thus, it became apparent to both the pro-trade and anti-trade groups (countries) that for ‘trade’ or ‘conservation’ to be sustainable, illegal wildlife trade should be managed. The transboundary migratory nature of wildlife also led to complications in its management and thus a transboundary cooperation arrangement in enforcement activities seemed to be the only way forward in addressing poaching and illegal trade in Africa.

(c) Illegal trade and link to organised crime: As the level of poaching increased in Africa, so did its link with other major organised crime network. There were newspaper reports on how warlords in DRC were benefitting from the trade. A series of recent UN reports (UNEP-MONUSCO-OSESG (2015)) provides some validity to this claim.

(a) Technological Development: When anti-poaching efforts took dominance in Africa, poachers and illegal traders resorted to the use of advanced technology (GPS-based tracking systems, drones, helicopters and automatic weapons) either to increase their effective ‘catch’ or to evade being caught. The global conservation efforts then took a new turn, where several technological innovations were called upon to address the problem (Force (2014); Marks (2014)). As a counter-measure, African countries also required map-based technologies to identify the hotspots of poaching and trafficking and similar technologies to capture the poachers. The use of technology though has become a clear standpoint for addressing wildlife crime in Africa, its use and application (especially the use of drones) has so far been limited due to its social and political implications (Sands (2003)).

Public Opinion

A notable shift in global public opinion also took place soon after CITES COP 15 in 2010 when it was noted that elephant and rhino poaching in Africa was related to the needs of the markets in key Asian countries. There was a wide range of media coverage on elephant poaching and its links to terrorism and arms trade. Stories about elephant and rhino poaching became headlines in newspapers and television media around the world. Illegal wildlife trade was also considered as a security problem; just not that it would deplete the wildlife population, but also because of fears about on how it could finance armed groups in Africa. The widespread public opinion also led to pro-trade countries in Asia, especially China, to raise concerns about illegal trade in Africa.

[6] News report on the threat of sanctions to Kenya, Uganda and Tanzania over illegal ivory trade. http://www.theeastafrican.co.ke/news/East-Africa-joins-the-infamous-Gang-of-Eight-over-poaching/-/2558/1943130/-/26u3yf/-/index.html

All the above-mentioned factors brought significant changes in the governing coalition; these changes are described in the next section.

Changes in Systemic Governing Coalitions

Although soon after CITES COP 15 in 2010, there was a wide range of publicity and willingness to address wildlife crime, the high level of poaching in Africa also led to serious mistrust among major bilateral donors towards sub-Saharan African Governments. This prompted western donors to channel their funding through international agencies and NGOs. Some argue that NGOs received substantial financial benefit from the crisis (Edge (2015)). But for NGOs, gaining funds was not sufficient for operationalising their pledges. They also required the governmental support to accomplish their goals. On the other hand, despite the fact that African Governments were not favourable to NGO involvement, they were aware that it would be difficult to move forward without the financial and technical support of NGOs. This particular 'interdependent' characteristic or 'marriage of convenience' is also clear from the empirical study carried out in sub-Saharan Africa where the governments themselves expressed their preference to work with 'certain' international NGOs (see Chap. 4, Section 4.5.2) whom they considerable more 'knowledgeable'. Such a phenomenon seems to be common in Africa as can be observed from previous studies. For example, Feldman (2012) recognised that channelling funds through NGOs has, in fact, reduced the conventional powers of African bureaucrats, especially regarding their role in managing natural resources, information flows and other processes. In some cases, bilateral donor funds were used to put government officials on to the payroll of international NGOs. This 'New Age' of environmental governance, and as mentioned by Mol (2006), has thus transformed information generation, processing, transmission and use into the fundamental (re) sources of power factions and transformation. Therefore, in contrast to the Asia case, for funding purposes, African countries had to co-opt with NGO visions as they remained the channel through which financial resources flowed into sub-Saharan African countries for issues related to wildlife conservation.

The involvement of NGOs in national enforcement-related activities and the reports they generated at times caused concern for CITES Secretariat, where CITES decisions were blamed, partly, for the rise in illegal ivory trade in Africa. Noting this trend, in 2010, CITES cautioned parties on involving NGOs in enforcement-related matters (see comments[7] by CITES).

[7] In recent years, non-governmental organizations (NGOs) have begun to play an increasingly significant role in the delivery of enforcement-related training and in developing enforcement strategies at national and sub-regional levels. However, the Secretariat has noted examples where government agencies appear to have almost abdicated their statutory and constitutional role to NGOs. In some parts of the world, NGOs appear to gain access to enforcement-related information in a manner that may be legally questionable. Several NGOs have attempted to establish themselves as communication channels between national enforcement agencies and international bodies. The Secretariat recognizes that the NGO community has a vital role to play in supporting wildlife law enforcement efforts. It is conscious that Resolution Conf. 11.3 (Rev. CoP14) specifically encourages Parties to work with NGOs with regard to enforcement matters. However, the

As a consequence of the above-mentioned factors, a shift in approach emerged that later led to a consensus between some of the pro- and anti-trade sub-Saharan African countries for tackling illegal wildlife trade. The influence of this consensus on coalition formation is further highlighted in Sect. 5.3.1.3.

Policy Decisions and Impacts from Other Policy Subsystems
Though the wide scale public opinion elicited commitment to save African wildlife, combating wildlife crime required large-scale investment. The financial crisis affected contribution to African governments by US and European Governments—both in terms of the amount and the way it was channelled. This meant high commitment but less financial resources for the subsystem actors to advance their goals. All these led African countries to seek new ways to address the challenges in transboundary illegal wildlife crime.

5.3.1.3 Long-Term Coalition Opportunity Structure

Earlier it was mentioned that Africa remains as one of the major suppliers of wildlife products to Asia, Europe and North America. This had also created a long-term partnership between governments and businesses in Africa and the rest of the world. The monopoly on the trade of certain species, in fact, divided the African continent into pro-trade and anti-trade nations which then led to cleavages or voting blocs on decisions related to those species (detailed further in sect. 5.3.1.5).

5.3.1.4 Short-Term Constraints and Resources of Subsystem Actors

The global financial crisis did constrain the resources and activity of the various actors involved in addressing conservation and trade issues and, a primary reason why African countries had to co-opt with NGO visions. Resource shortages also triggered temporary partnerships among the sub-Saharan African countries on joint operations (where resources are shared) in addressing wildlife crime.

5.3.1.5 Coalitions, Policy Beliefs and Resource Availability

In Chap. 4, I mentioned that conflicting beliefs in the wildlife–trade policy subsystem are primarily based on prioritising aspects of nature conservation and trade. The study that was carried out in Kenya, Uganda and Tanzania (all member states of LATF) revealed these distinct core beliefs on: conservation issues, which reflected ecocentrism and protectionism; utilitarian issues, sometimes taking the pretext of sustainable use; and the core regulatory or hierarchical approach, often reflected by

Secretariat also believes that, in an increasing number of instances, the correct balance is not being struck. It is aware that other international organizations share its concerns (CITES, 2010a).

government authorities. For a long time, these conflicting policy beliefs influenced the sharing of information between LATF member states on the key polarising issues, including ivory trade. As a result, and similar to what was happening with the databases of INTERPOL and CITES, LATF member states were selective in sending complete information details on wildlife crime to an MS Access database[8] managed by LATF. Pro-trade nations within LATF were wary of sharing their information on wildlife crime with LATF headquarters as it was based in Kenya, a country that was seeking a total ban on ivory trade. In 2005, when LATF appeared to be falling apart, LATF Parties undertook an evaluation and assessment of the implementation and impact of the agreement during its first decade of existence (1994–2004). As a result of this evaluation, a set of key recommendations was outlined for its improvement (UNEP, 2005). The evaluation report and the recommendations made were later adopted at a meeting of the Lusaka Agreement Governing Council (UNEP, 2005). Based on the decision of the Governing Council, the Parties in collaboration with UNEP and other partners developed a long-term strategic plan of action for the implementation of the agreement on a short-, medium- and long-term basis. The strategic plan was reviewed in a regional workshop in July 2005 where three major recommendations[9] were made as indicated below:

(i) Develop a transboundary database to share wildlife crime information between the member states;
(ii) Undertake strategic assessment of the Lusaka Agreement activities;
(iii) Promote international cooperation with donors and international agencies.

However, a true consensus between the opposing coalitions emerged only after CITES COP 15, in 2010.

Strategy

Soon after CITES COP 15, LATF initiated its preparation for the development of a transboundary enforcement information system. The timing of this preparation coincided with the development of two transboundary databases—WEMS and EU-TWIX. WEMS and EU-TWIX have a lot of commonalities. They monitor transboundary illegal wildlife trade and have the potential to highlight trends and threats on enforcement matters. WEMS and EU-TWIX were conceptualised by recognised international institutions and their purpose remained the same. However,

---

[8]The first database developed by LATF was in the format of Microsoft Access, which generated reports on: recovered and confiscated weapons; poaching statistics; wildlife product statistics/seizures (animal trophies); live wildlife specimen statistics/reports/seizures; and live wild flora product statistics reports/seizures. The format for these five reporting forms was designed in-house by LATF with the assistance of a database specialist from the Kenya Wildlife Service. The reports were submitted via email, fax and post by the member states and then incorporated into the MS Access database (UNEP, 2005). However, mistrust between member states, a shortage of the relevant staff and skills required to develop and maintain the database remained a big challenge to the fulfilment of this operation.

[9]Extracted from the UNEP evaluation report: *A Decade of Regional Wildlife Law Enforcement: The Case of the Lusaka Agreement* (UNEP, 2005).

the major difference was in the design structure which apparently depended on the coalition of partners who had framed it. While the coalition supporting EU-TWIX comprised of CITES and the European Union was supported by international NGOs such as TRAFFIC, WEMS was driven entirely by a group of researchers at UNU (by this time, UNU had ended its formal collaboration with ACA). Even though CITES had approved EU-TWIX as a potential transboundary enforcement system (CITES, 2009), with the plan for it to be expanded as a worldwide system called 'Global-Twix', the Secretariat later admitted that it had no capacity to take on such work and none of the organisations were ready to do it for free (see Chap. 2, Section 2.4).

Decision Outcome

Now, returning to the reasons for the adoption of WEMS in Africa, it is clear that the main reason for the adoption was due to the existence of a favourable or collaborative policy subsystem at the regional level in sub-Saharan Africa, where the Governing Council of LATF had already approved a decision to implement a suitable transboundary information system before WEMS was even presented to it. The specific aspects within the policy subsystem that favoured WEMS are detailed as follows:

1. Members who supported WEMS were in positions of formal authority: The will to undertake WEMS came from the Director of LATF and was supported by the Governing Council which was composed of Ministers as representatives from the member states. Since the LATF Governing Council had already decided to implement a transboundary information system, it was just a matter of choosing the appropriate one. As WEMS was designed and managed by a UN organisation, it gained the trust of the Governing Council to approve the implementation of WEMS in Africa.
2. Favourable multi-stakeholder (UN, Governments and NGOs) opinion: Another important reason for the adoption was due to favourable multi-stakeholder consensus on addressing wildlife crime. The decision to host a transboundary enforcement monitoring system had been on the agenda of CITES meetings (CITES, 2004, 2007b, 2011a, 2011b) for a while, and several regional bodies, including ASEAN and LATF, had earlier shown interest in developing one. Several media articles had been critical, especially on China and African countries, for not doing enough to curb wildlife crime. Therefore, there was a general consensus on having a system that could monitor illegal trade across borders.
3. Financial resources: WEMS was provided for free to LATF. International Fund for Animal Welfare (IFAW, an NGO) and UNEP were also supportive and provided the resources to carry out the training programme and workshop required for implementing WEMS. So there was no shortage of resources in getting the implementation process working.
4. Brokerage: The Director of the Lusaka Agreement showed leadership in bridging consensus among the various parties in adopting WEMS. This won the trust of the Lusaka Agreement Governing Council to implement his decision. LATF had also partnered with other reputed research institutions, including University

of Washington, on DNA analysis, and peer-reviewed scientific articles (Wasser et al., 2008; Wasser et al., 2007) emerged from their cooperation. In 2013, the Lusaka Agreement later won the CITES commendation award (CITES, 2013e) for their work in addressing wildlife crime.

5. Timing: The financial crisis in Europe and in North America had reduced donor funding for conservation and enforcement matters (until 2011). On the other hand, increased demand for wildlife products in China and other Asian countries had resulted in an aggravated situation of wildlife crime, especially with regard to high-value products such as rhino horn and elephant ivory from Africa. In the absence of a comprehensive database for monitoring such crimes, it was difficult to provide conclusive evidence on the perpetrators and on issues related to non-compliance to the CITES Agreement by both the 'producing' and 'consuming' countries. The analytical capability within WEMS could help map out the producer and consumer countries (Chandran et al., 2013).

Policy-Oriented Learning

One key factor which enabled the adoption and implementation of WEMS in sub-Saharan Africa was the alteration of thoughts with regard to enforcement information-sharing between pro-trade and anti-trade sub-Saharan African countries. For instance, sub-Saharan African countries have never been allowed to participate in the one-off sale of ivory since the trade ban on ivory. In 2010, when South Africa, Botswana and Namibia were given permission to trade their stockpiles of ivory, CITES put on hold the trade from rest of the elephant range states, citing poor enforcement efforts. Such a decision has prompted the pro- and anti-trade countries in sub-Saharan Africa to join hands to work on enhancing enforcement and compliance with CITES through transboundary cooperation. It should be noted that the decision to implement WEMS came after the CITES COP15 meeting, where a consensus agreement to continue the ivory ban was already in place. Meaning—sharing information with a regional body was not going to affect international decisions on ivory trade. This triggered a general agreement among the pro-trade and anti-trade factions in sub-Saharan Africa in addressing illegal wildlife crime through cross-border information-sharing. It should however be noted that a decision to share enforcement information doesn't mean a position change by the countries on key wildlife issues; rather, it is more a mechanism to increase the knowledge of the patterns of illegal trade and the factors influencing it. By doing so, they can make revisions in the policy problem and thereby influence decisions at CITES meetings.

A second aspect of policy learning among the Lusaka Agreement coalitions can be considered in their shift to work with scientific institutions instead of NGOs in order to enhance the data collection and sharing mechanism. Such an approach also reduced their dependence on NGOs in information collection and analysis process.

5.3.2 Analysis of Adoption of WEMS-Africa

When analysing the reasons for the adoption of WEMS in Africa, it can be concluded that, during the time of implementing of WEMS in the pilot countries, the regional policy subsystem was relatively stable. The evaluation report (UNEP, 2005) on the Lusaka Agreement had already stressed the need for a cross-border information system and, most importantly, the Governing Council of the Lusaka Agreement (even though Parties constituted different policy beliefs on wildlife trade) had approved the decision. By the time LATF approached UNU, it was simply a choice between the different information systems (secondary beliefs) that they had to adopt for implementation. The institutional rules, resource allocation and selection of the database all favoured WEMS due to the following reasons:

- It was free of cost which required minimum resource allocation;
- Funding was assured by the International Fund for Animal Welfare (IFAW).[10]

Now here is the distinction between the policy subsystem at the global level and at the regional level:

- At the global level, the initial discussions were to have a global database but this failed to materialise as there was no consensus among CITES and other relevant agencies to take the responsibility for developing and hosting a system (see Sect. 2.4 and Sect. 5.3.2).
- While this discussion was going on at CITES, the WEMS system was already being implemented in some LATF countries. When WEMS was presented at the CITES COP16 meeting in Bangkok, it had recorded a total of 164 cases of wildlife crime from the four pilot countries participating in the initiative (Chandran et al., 2013). At the time of finalising this book, WEMS has recorded nearly 534 cases (LATF, 2015) and is considered as working model for transboundary enforcement information-sharing on wildlife crime between Lusaka Agreement member states (LATF, 2014, 2015).

This indicates that at the regional level, the policy actor has a greater opportunity to develop change within the policy subsystem and can be seen as a reason why a regional system may be more appropriate than a global one. Another reason is that the regional policy actors have more opportunity to understand the technical aspects of the topic (enforcement information-sharing), its context (sharing information) and its geographical application. However, it should be noted that regional policy subsystems are not immune to the effects of the larger political environment in which they operate. They are also affected by changes at the global level.

[10] It should be noted that IFAW had an interest in the continuation of WEMS, primarily because they considered WEMS as a tool that could support enforcement efforts in Africa. IFAW also funds INTERPOL activities. Please see IFAW press release, *New and improved system to combat illegal wildlife trade unveiled* http://www.ifaw.org/united-states/node/30721

5.4 Conclusion

From the case studies of the objection of WEMS in Asia and its acceptance in Africa, it is clear that the dynamic features of the acceptance of transboundary information systems are not only based on technological brilliance but also on the interactions between the actors within the policy subsystem and the resources of the actors.

In an attempt to understand the underpinning policy factors that led to the objection to WEMS in Asia by CITES and the processes that led to its subsequent adoption in Africa, the study revealed a major difference between the policy subsystems in Asia and Africa. WEMS-Asia was developed within an adversarial policy subsystem where the coalitions were in conflict due to their conflictive policy beliefs, whereas in the case of WEMS-Africa, even though the coalitions were in disagreement over their core policy beliefs, each coalition did have their own reasons for forming a collaborative decision on having a transboundary information system. Also, during the time of WEMS-Africa implementation, wildlife crime was high on the agenda.

Apart from the interactions within the policy subsystem, a common factor which influenced both the Asia and Africa cases was the dynamic influence caused by external factors. As wildlife–trade policy subsystems are complex systems which remain nested and interdependent on other subsystems like trade policy, law enforcement and finance, the factors causing changes within these systems also have an effect on the decisions made within the wildlife policy subsystem. These dynamic changes are unpredictable and can cause shifts in the policy approach.

Several literatures define the role of a policy broker as that of a consensus builder among the conflicting coalitions. However, in this study, the CITES Secretariat, the key policy broker, was weakened by a shortage of resources needed to maintain its functions, and therefore had to resort to a 'policy bent' (this policy bent has also been observed in similar cases by Sabatier (1988) and Smith (2000) to get the support of the resource-rich factions.

A notable feature of the study is the alteration of thoughts or behavioural intentions of the coalitions resulting from experience. UNU scientists disassociated themselves from ACA to form a new alliance with government coalitions and, in sub-Saharan Africa, the opposition coalitions joined hands to share transboundary enforcement information in spite of disagreement on core values. Hence, *policy-oriented learning* is an important feature within this study which enabled the different sets of actors to understand the situation and further enhance their policy objectives. Although broad participation is considered a key necessity in political systems (as seen during CITES COP meetings) where expertise is diffused among government and non-government actors as well as scientists and consultants, in addressing a complex problem, this broad participation also produces disagreement and raises legitimacy issues (see Dryzek, 2001). This means that the addressing of complex problems requires closed-door and frank communications between adversarial groups where the groups can then define a moderate path for solving the problem. A practical experiment on achieving a consensus on WEMS is outlined in the next chapter.

Finally, it should be noted that WEMS is showcased as a scientific tool in many literatures (Chandran et al., 2011; Estevez & Janowski, 2013; Kretser et al., 2014). However, social scientists would interpret it as an artefact which was developed to endorse the beliefs of certain factions. The true value of WEMS can be achieved only if it acts as a consensus object (boundary object) between the conflicting policy beliefs while at the same time provides the necessary technical input that addresses the shared values of the coalitions. In other words, for WEMS to be successful, it should be of use to the coalitions within the policy subsystem, acting as a channel for peaceful communication.

Chapter 6
Bridging Multiple Social Worlds: Boundary Work in the Adoption of WEMS

6.1 Introduction

From the previous chapter, it is clear that the complexities in the adoption and implementation of a transboundary information system are influenced by several factors, including the disparities in policy beliefs and resources of multiple institutions operating at different scales. Though WEMS was successfully implemented as a regional database in sub-Saharan Africa, it was still not recognised by global enforcement bodies. From the next paragraph onwards, this chapter explains on 'how WEMS was made acceptable among the various global stakeholders within the wildlife policy subsystems'.

After CITES COP 16 in Bangkok, it was clear that the technical value and the successful implementation of WEMS were not a criterion for the adoption of WEMS at a global level. What is required for WEMS to be functional at a global level is a new framework that can fit both the global and regional contexts. The question for UNU scientists and project managers of WEMS therefore was not on 'whether' to move ahead with the initiative but rather on 'how' to modify its programme design and practices in ways that will help to realise its acceptance and implementation.

But this task was not so easy. The very reason that a lead institution like CITES had not endorsed the 'solution for enforcement and compliance' (as mentioned by Chandran et al., 2011) made the task of a rather irrelevant institution (in the context of the wildlife policy process) like UNU more difficult. Secondly, at the CITES COP 16 meeting held in March 2014, the convention had already decided that the enforcement information collection will be done through INTERPOL and the World Customs Organisation (WCO). It decried WEMS on the grounds that WEMS may duplicate the efforts of existing global enforcement bodies that are already entitled to collect, compile and share information.

We already know that this is not the first time such a decision has been made by the CITES Secretariat. In 1994, when sharing enforcement information was not

R. Chandran, *The Wildlife Enforcement Monitoring System*, Professional Practice in Governance and Public Organizations,
https://doi.org/10.1007/978-3-031-97084-9_6

effective, CITES recommended that the existing global enforcement agencies take the lead (see Chap. 2, Section 2.4). The only difference between now and 1994 is that there was no full-time officer dealing with wildlife crime at INTERPOL's Lyon office. Now, with dedicated personnel, a full-fledged programme (which apparently had started from an NGO contribution) on environmental crime had been established at Lyon (INTERPOL, 2014). In addition, the ICCWC consortium (CITES, 2010b) to which CITES, INTERPOL and WCO belong, has become a powerful decision-making body directing and implementing global wildlife enforcement policies.

So, what did all this mean for WEMS? It meant that if WEMS desired to be considered or endorsed by the global supranational institutions mentioned earlier, it needed to discipline itself to the structures and norms of the global enforcement agencies. Such a move would provide a 'global' brand for WEMS. However, this does not mean that a global endorsement will make WEMS functional at all levels; rather, it will still depend on the regional and national agencies to comply with the requirement of information input into the database. On the other hand, an objection on WEMS by the supranational institutions can lead to a slowing down or even objection of the system at regional and national levels, just because national governments may not want to waste their time and resources on a tool which is not recognised or licensed as a compliance monitoring tool.

In other words, for WEMS to be functional, it will have to meet the requirements of both the global and regional wildlife law enforcement agencies. It also means, WEMS cannot work in isolation as a science-based decision tool with its functionality solely dependent on what it can bring to scientific research rather than what form it will take in addressing the concerns of law enforcement.

The other challenge for accepting WEMS by the global enforcement agencies was that INTERPOL and WCO had already developed databases on monitoring wildlife crime and other forms of environmental crime. For instance, the information for WCO comes through a structured process of information collection channelled via its Regional Intelligence Liaison Office, or RILOS.[1] In addition, it operates the National Customs Enforcement Network (nCEN)—a system developed by WCO to assist customs administrations with the collection and storage of law enforcement information at the national level, with the additional capability to exchange this information with regional and international levels. Similarly, INTERPOL has a secure information-sharing protocol, which allows National Central Bureaus (NCB)[2] to provide crime (all forms of crime) information to its

[1] WCO's role in combating environmental crime:
http://www.wcoomd.org/en/topics/enforcement-and-compliance/activities-and-programmes/ep_environmental_crime/wco_and_envrionmental_crime.aspx

[2] All INTERPOL member countries have a National Central Bureau (NCB), which links the national police with INTERPOL's global network. It is typically a division of the national police agency or investigation service and serves as the contact point for all INTERPOL activities in the field. http://www.interpol.int/About-INTERPOL/Structure-and-governance/National-Central-Bureaus

headquarters in Lyon. The formation of INTERPOL's Environmental Crime Programme has helped facilitate NCBs to send ecomessages to INTERPOL office—something that is mandated by the CITES Convention and is part of its global compliance procedure (CITES, 2007a).

However, it should be noted that the formation of powerful institutions at the global level will not guarantee a smooth flow of information from a national or a regional level to a global level. The bad news is that countries do not supply complete and often relevant information on wildlife crime into a global system—an indication of how non-disclosure is used as a tool of power at the global, regional and national levels.

Although the proclivity of individual countries for non-disclosure remains bad news to global enforcement agencies, the sharing of information across borders using WEMS turned out to be the *raison d'être* for WEMS. If WEMS can create some form of working arrangement between regional enforcement networks and the global enforcement agencies, then it will be the chief selling point for its functioning/existence.

In policy terms, the acceptance of WEMS was solely determined by its capacity to act as a 'boundary object' in which it should first qualify as a string that links multiple organisations set up in different geographical and political contexts (described later in the chapter through the concept of social worlds). It does not matter whether WEMS could come close to doing justice to what it was earlier designed for in 2005—i.e. open-access information. What mattered more was the role WEMS could play in increasing the linkage and trust between the multiple social worlds. To analyse the process of turning WEMS into a 'boundary object', it is important to shift from a long-term, meso-level policy dynamics (like in WEMS-Asia and WEMS-Africa) to a short-term micro-management analysis of WEMS as a boundary object. The next section enlightens on the theoretical concepts that were applied practically in turning WEMS into a potentially successful boundary object.

6.2 Pivotal Theoretical Concepts

Linking multiple social worlds is complex and needs thorough investigation of the interactions that take place and of the objects that people create and use when crossing the boundaries of different social worlds and communities (Worrall, 2012). This requires some historical understanding of the problem and how much consensus has been achieved in addressing the problem. There are two theoretical concepts that can be applied in this context— Problem Structuring and Boundary Object theory.

6.2.1 Problem (Re)structuring

Hoppe (2011), in his book, *The Governance of Problems,* describes the process of *problem structuring (or solving)* as a mechanism to bring a relative consensus on addressing the issue in question. He uses *problem structure* as a core variable and defines a *structured problem* as one where actors are close to agreement on the norms and values at stake, and the required knowledge is available and certain (see Fig. 6.1). A structured problem can be solved by outsourcing it to an expert community or technocrats. Whereas in an unstructured problem, there is a deadlock in the decision-making process due to agonistic politics. Defining the theoretical underpinnings within the context of WEMS, one can argue that in the initial stages (WEMS-Asia), UNU considered transboundary enforcement information-sharing as a structured problem where they opted for a technical solution. However, later they realised that the issue in context was complex or an unstructured problem.

While solving a complex policy problem, it is first important to create a structure for the problem prior to subsequent evaluation of the different action options. The dynamic process of structuring a decision problem involves the specification of options, attributes for evaluating options, and states of nature that may occur, with repeated circling back in the process to revise or augment the structure. However, in many policy problems, where tasks are usually ill defined, i.e. the options, attributes, outcomes and states of nature are not yet specified, the process of structuring is complex. One reason for this complexity is that policy forums are used as an arena for conflict over the concepts used in framing political judgements on social problems (Hoppe, 2011). In other words, a venue (for example, COP meetings) which is designed for problem solution or 'structuring' becomes a battleground of views of beliefs, resource allocation and power.

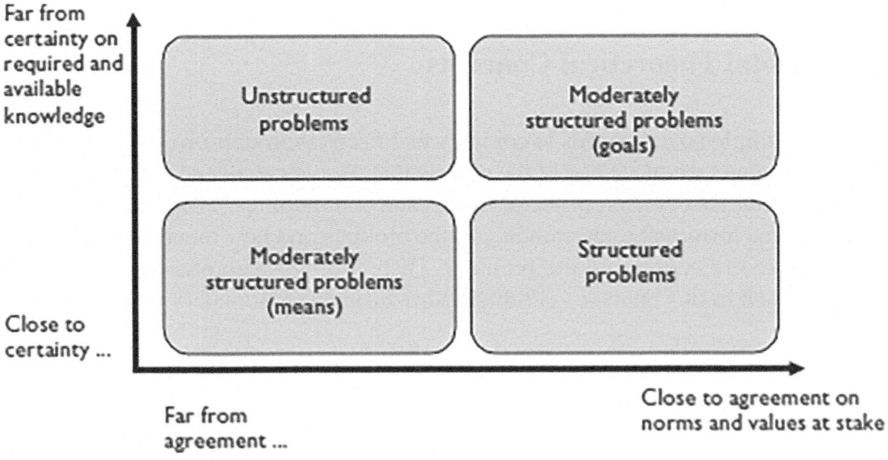

Fig. 6.1 Simple typology of problem structures (Hoppe, 2011)

As a process to identify a solution, Hisschemöller and Hoppe (1995) classify problem structuring by emphasising two important intermediary processes based on 'means' and 'ends'—considered as steps in the problem-solving processes from an unstructured to a structured problem. Referring to these intermediary processes as moderately structured problems, the first one is mainly characterised by the cost-benefit bias (see Fig. 6.1). The policy strategy linked to this problem—where there is a general consensus about the main thrust of policy, even though policy goals may not be as clearly defined as in the structured problem situation—is referred to as negotiation (Hisschemöller & Hoppe, 1995). As conflict revolves around the 'means' to reach the policy goal more effectively and efficiently, negotiations are regarded as institutionalised short-term mechanisms for resolving conflict (Hisschemöller & Hoppe, 1995). In the longer run, applied research focusing on effectiveness and/or efficiency of alternative means is also an appropriate strategy.

In the second scenario of a moderately structured problem, the dispute is about discordant values or 'rights'. This conflict is mitigated by incorporating into a compromise the values most relevant to the conflicting Parties (Fig. 6.1). Hisschemöller and Hoppe refer to the policy strategy that belongs to this type of problem as 'accommodation'.

Returning to the context of WEMS, one can easily observe the patterns of problem structuring during the 10-year course of its development and implementation (for details, see Table 6.1, in Sect. 6.3). Though there were several cost consequences to the delay in the adoption of WEMS, the multiple stages the project went through was helpful in identifying certain patterns of facts and values, especially between the boundaries of institutions of science and institutions of politics and between politics and administration.

It should be noted that 'values', 'means and 'ends' remain the core characteristic of any organisation—inscribed within them when they were created. Changing them will be difficult and is not the way to go ahead in finding a consensus. The way to move forward is to identify mechanisms that can bring consensus or congruence among values, means and ends. In other words, while means and ends have been considered ways of structuring an ill-defined problem, structuring can never be complete until it is constituted of one or several 'boundary objects' that can help in facilitating interactions, translations and coherence across different conflicting means and values. In practical boundary work, the fact–value distinction is continuously appealed to and sometimes renegotiated as a basis for demarcation and coordination of activities of organisations (Hoppe, 2011).

The key elements of the boundary object theory are outlined in the next section.

6.2.2 *Key Elements of Boundary Object Theory*

Boundary object theory was originally developed by Star and Griesemer (1989) and has been applied (Levina & Vaast, 2005) and used (Bechky, 2003) to study the interactions that take place within different policy conflicting scenarios and the objects

Table 6.1 Problem structuring in WEMS: sequence of events

Organisations Conflicts within WEMS		Type of Conflict	Problem	Type of solution (Moderately structured—MS1-Negotiation or MS2-Accommodation)
Stage 1–2005-07				
Type of organisation-1	Type of organisation-2			
CITES –UN policy making body (A)	UNU—Research arm (B)	Activities of (B) causes conflicts in the decisions of (A)	Organisation B believes in openness, fairness and transparency - NGOs should be involved in bringing out the truth on wildlife crime. Organisation A believes in a closed-door policy on enforcement information-sharing—NGOs should be limited only to capacity development activities	B adopts a's idea—WEMS proceedstostage2–**Process:** Negotiation
Stage 2–2007 -10				
CITES –UN policymaking body (A)	UNU—Research arm (B)	CITES COP 15 fails to endorse WEMS as it was not tested in the pilot phase	CITES recommends EU-TWIX as a model for global implementation. While WEMS had to be tested at a regional and national level	B tests WEMS as a pilot phase in Africa and brings the results back to A in 2013. EU-TWIX fails to be adopted globally **Process: Applied research**

(continued)

Table 6.1 (continued)

Organisations Conflicts within WEMS		Type of Conflict	Problem	Type of solution (Moderately structured—MS1-Negotiation or MS2-Accommodation)
Stage 3–2010-13				
CITES –UN policy making body (A)	UNU—Research arm (B)	CITES secretariat suggested that WEMS should not be promoted as it might overlap with, duplicate, or divert attention and resources from, efforts to improve the submission of enforcement information through the INTERPOL and WCO information systems.	CITES secretariat and its ICCWC partners, INTERPOL and WCO, are committed to encourage wider and better use of the existing global information systems, which have been operated by INTERPOL and WCO for many years and which are supported by a network of dedicated national and regional centres.	**Process:** Negotiation

that people create and use in the context of crossing and relating the boundaries of different social worlds (shown later in the section) and communities. The theory uses and adapts concepts that were originally conceived by other researchers, including Akrich, Latour, Callon (Akrich et al., 2002) and Law (Callon & Law, 1982), while introducing its own concepts—most notably that of the boundary objects themselves. The theory then relates these concepts to the role of boundary objects in facilitating interactions, translations and coherence across social worlds.

There are four core concepts used in boundary object theory, namely: social worlds (as mentioned by Strauss & Fagerhaugh, 1997; Star & Griesemer, 1989), interessement and translation, boundary objects and coherence/convergence (which are given separate names but should be considered essentially the same concept). These concepts are outlined in detail as follows:

Strauss and Fagerhaugh (1997) defined the theoretical concept of social worlds as featuring: (a) one or more primary activities; and (b) locations where activities occur. They further describe how it is constituted of secondary concepts or features which include a technology that allows activities to be carried out that helps the establishment of social worlds and organisations that further that social worlds activities. Strauss and Fagerhaugh (1997) also believed the study of such worlds, at any scale, should not be confined to their discourses but also consider their activities, memberships, sites, technologies and organisations. Taking an ecological approach to social worlds, Star and Griesemer (1989) reflected further on the concept and describe how social worlds represent the views of a group or a culture. They show how people enter an association or a network with their given identities and perspectives, and why it is difficult (also unnecessary) to expect them to throw away their original identities and perspectives simply to embrace the newly emerging, collective ones. According to them, people's newly emerging collective identity and perspective may build on the fact that social actors understand that their given identities, and perspectives are included, considered, and respected. This places social worlds in the framework of national and regional ideologies and how organisations (at global, regional and national levels) are identities that reflect the constellation of beliefs and ideologies of actors administering the organisation.

Boundary object theory is also based on the derived concepts of *translation*, which is adapted for use in the theory (Worrall, 2012). Translation itself is defined in the context of multiple social worlds as the task of reconciling the meanings of objects, methods and concepts across these worlds so that people can work together. As mentioned earlier, in practical boundary work (which partially gives a different account of translation), the fact–value distinction is renegotiated through demarcation and coordination (described in detail in Sect. 6.5.1) of activities of organisations (Hoppe, 2011).

Boundary objects are entities that allow heterogeneous actors to develop trust and form stable (if perhaps, transitory) working relationships (Worrall, 2012). Boundary objects can exist between a single pair of actors or among multiple Parties, each with their own opinions and perspectives. Boundary objects are 'plastic' enough to adapt to changing needs (Star & Griesemer, 1989). For example, in the case of WEMS, the information system itself or its derivatives like maps and data should be able to connect and be useful to all the stakeholders (belonging to different social worlds) in some way or other. In practical terms, Boundary objects can also be classified as 'working arrangements', adjusted as needed. They are not imposed by one community or by appeal to outside standards (Bowker & Star, 1999). Boundary objects should also satisfy different concerns simultaneously (Clark et al., 2011). In other words, Boundary objects allow different groups to work together without consensus (Worrall, 2012). The interpretation flexibility, the dynamics between ill-structured and more tailored local uses, and the satisfying of information and work requirements elucidate Boundary objects' crucial role (even if only temporarily) in solving a complex or unstructured problem by encapsulating needs at both local and collective levels (Worrall, 2012). In the case of WEMS, the Boundary object should overlap boundaries between multiple social worlds (CITES, INTERPOL, WCO) and levels (local, regional, global) for it to be practically functional.

Coherence and Convergence is another important concept of boundary object theory. According to Star and Griesemer (1989), Coherence is the degree of consistency between different Translations and Social Worlds. For example, conveying or translating the work of scientists to non-scientists (or the other way around) requires practical commonality or 'meaning capturing' among the two actors. Boundary objects play a critical role in developing and maintaining coherence across intersecting social worlds. Convergence, on the other hand, considers how well the tools, systems, interfaces and devices for storing, tracking, displaying and retrieving information conceptualised as information artefacts are fitted to communities of users that create and work with them (Worrall, 2012).

6.2.3 Practical Application of Boundary Object Theory

The original application of Boundary object theory was primarily restricted to scholars mainly in the field of science and technology studies. Recently, Boundary object theory has arguably become a grand theory through its wide application in other disciplines, as can be seen in the work of practitioners as well as scientists (Clark et al., 2011; Emad & Roth, 2009; Levina & Vaast, 2005; Mcknight, 2007). Of the recent works, Clark et al. (2011) apply Boundary work for sustainable development studies, and Mcknight (2007) uses the Boundary theory to enlighten the role of multiple boundary objects in finding a solution in a high-conflict arena. The works of Clark et al. and Mcknight were specifically relevant in bringing out WEMS as a successful Boundary object.

The next sections will analyse in detail how WEMS was repositioned from an unwelcome duplication of existing wildlife monitoring and information systems to a Boundary object enabling their convergence for global policy purposes.

6.3 Unearthing the Differences

Realising the need for getting WEMS functional at both ends (national and global), in 2013, UNU scientists further carried out semi-structured interviews with the various stakeholders from enforcement agencies, including INTERPOL, World Customs Organisation (WCO) and United Nations Office for Drugs and Crime (UNODC)—the three main actors who were engaged in wildlife crime enforcement information-sharing and analysis and also part of the newly formed ICCWC consortium.

The interviews revealed certain gaps in inter-agency coordination among enforcement agencies at national and international levels, which are described below:

6.3.1 Conflicting Mandates

First, the mandates of enforcement agencies like the United Nations Office for Drugs and Crime (UNODC), World Customs Organisation (WCO) and INTERPOL are not exclusively for monitoring wildlife crime. It is one part of their job among the myriad other more important matters that need to be looked into, like drugs, weapons and terrorism. Though INTERPOL had established the environmental crime programme, the fund for the programme had to be resourced externally.

Secondly, and probably the most important concern is that, there remain 'stove pipes' in inter-agency information-sharing at the national level originating from the specific mandate each organisation (Police, Customs and Wildlife Divisions) has in carrying out its work independently. For example, at Customs, the main purpose of information is not just to identify repeat perpetrators or offenders, but to examine how the concealment of contrabands is carried out and to identify the origin, transit and destination of the contrabands. For this reason, World Customs Agency does not usually hold nominal information (names of people) on wildlife crime. This task is left to the national customs agency where they compile nominal information into the National Customs Enforcement Network (nCEN). In other words, a national customs agency has no mandate to share nominal information with international agencies, while nominal information is one of the key information types that police agencies need for their investigation.

INTERPOL requires the names and details of identity sent to their headquarters in Lyon to make sure the perpetrator is labelled as an international criminal. The National Central Bureau (NCB), which holds nominal information, on the other hand, decides which information to send to INTERPOL. This decision is apparently a State prerogative and cannot be interfered with by INTERPOL. Most countries have reservations in labelling a convicted perpetrator as an international criminal and this leads to gaps in information exchange and data within the global information systems.

Another revealing finding from the interviews was that agencies belonging to the same grid (international agencies, including UN) can come into conflict or turf war when it comes to the action of one agency influencing the functioning of the other, either in donor competition or duplication of work. A UN survey explains in broader terms how agencies compete for resources and power (UNDESA, 2012).

The other problem that was a cause of conflict between national and international enforcement agencies was about the purpose of information at the international level and the needs at the local level. Government agencies operating at national and local levels are more concerned about how national decisions can affect the local stakeholders—for instance, how a decision will influence the action of a business group or a local community. Hence, actions taken at a global scale with ramifications for the national stakeholders of a country can cause conflict. To prevent free-riding by States on enforcement matters, international organisations at times behave as autonomous sites of authority. As international organisations are independent from the State 'principles' who may have created them and, because of power owing to their legitimacy as a rational-legal authority, they can be in control over technical expertise and information (Barnett & Finnemore, 1999).

Other issues that emerged from the interviews were the different approaches institutions take based on their disciplinary expertise. For instance, enforcement agencies frame the problem of wildlife crime as a security issue, and enforcement information as a sacred tool, not to be shared lightly with academic or civil society partners. On the other hand, academic institutions need information on wildlife crime for recognising the importance, relevance and the scale of the problem at a global level and an approach in finding a scientific solution.

Some of these findings are not new. Gaps in inter-agency coordination on enforcement information-sharing in general are well documented by several authors (Chandran et al., 2011; Dawes et al., 2009). A recent example[3] comes from the Boston bombing incident in April 2013 where the failure in detection has been attributed to non-coordination between the enforcement agencies. As wildlife crime remains outside 'mainstream' crime and is a victimless crime (Wellsmith, 2011), it is quite obvious why its record-keeping may not be in the way that drug-trafficking, murder, rape or burglaries are. This partially shows why national governments do not have a systematic report on wildlife crime information. But this does not mean they are not bothered as can be seen in the next section.

6.3.2 Sovereignty

State sovereignty denotes the competence, independence and legal equality of states. The concept is normally used to encompass all matters in which each state is permitted by international law to decide and act without intrusions from other sovereign states. These matters include the choice of political, economic, social and cultural systems and the formulation of foreign policy. The scope of the freedom of choice of States in these matters is not unlimited; it depends on developments in international law (including agreements made voluntarily) and international relations.

Within the Charter of the UN, there is an explicit prohibition on the world organisation from interfering in the domestic affairs of member states. The Charter's most frequently cited provision, Article 2 (7), provides that '[n]othing contained in the present Charter shall authorise the United Nations to intervene in matters that are essentially within the domestic jurisdiction of any State or shall require the members to submit such matters to settlement under the present Charter' (UN, 1945).

In other words, sovereignty is a constitutional safeguard which allows States to challenge intervention by external actors in State matters. According to the sovereignty principle, the State remains the fundamental guarantor of information locally, as well as the building block for collectively ensuring international order.

The sovereignty card is often used by national enforcement agencies in Asia and Africa as a reason for not disclosing information to international agencies. For instance, national enforcement agencies are reluctant to share nominal data with international agencies unless the perpetrator is a transnational criminal. In the case of wildlife crime, and in most instances, poaching is done with the involvement of

[3] 'I'm very concerned that there still seem to be serious problems with sharing information, including critical investigative information', Senator Susan Collins, Republican of Maine, told reporters. 'That is troubling to me that this many years after the attacks on our country in 2001, that we still seem to have stovepipes that prevent information from being shared effectively, not only among agencies but also within the same agency, in one case'. http://www.nytimes.com/2013/04/24/us/politics/senators-say-case-indicates-that-problems-persist-in-agencies-data-sharing.html?_r=0

locals, and the transit of the contraband to the *entrepôt* (or port of export) is operated by notorious criminals (middlemen) holding State nationality. In either of the cases, both the poacher's and the middleman's operation remain within the jurisdiction of the State. For this reason, national enforcement agencies do not see the need for informing an international enforcement agency regarding crimes reported within its jurisdiction and prefer to deal with them themselves until the international context of the crime is proven. This justifies why reporting to an international body remains skewed or most often does not occur at all.

But international law enforcement agencies do not always attribute the skewed or missing reporting to claims of State sovereignty. In fact, at times, this leads to apprehension, suspicion and accusation from global enforcement bodies. A former CITES enforcement officer asserts the reason as follows: '*they (national enforcement agencies) don't supply it (information) because they can't be bothered to. They don't take wildlife crime seriously enough. Shameful, isn't it?'* (Sellar, 2013).

It cannot be denied that purposeful non-disclosure of information happens when the issue in context compromises the political bargaining power of the State as can be seen in the debates on ivory and rhino-horn trade (See Chap. 2, Section 2.2). One reason for this is the conflicting beliefs among different stakeholders within the wildlife policy subsystem. This is further described in detail in the next section.

6.3.3 Conflicting Beliefs

As mentioned earlier, one main reason for purposeful non-disclosure of information that constrains information-sharing is the polarisation based on differing wildlife policy beliefs among stakeholders at the national, regional and global levels. This polarisation also has a significant influence on how information is used and on how decisions are shaped during CITES meetings. Polarisation in beliefs is already outlined in detail in earlier chapters. The effect of polarised beliefs on information-sharing is also well documented by several researchers working on wildlife conservation and enforcement matters (Duffy, 2000; Gibson, 1999; Gockeritz et al., 2010; Reeve, 2014; Sukumar, 2003). WEMS itself has been a victim of conflicting beliefs and the mandates of different agencies (see Chandran et al., 2015).

6.3.4 Resource Crunch

The other important factor which stood against the adoption of WEMS after its implementation in Africa was the fear of shifting attention and resources away from global enforcement institutions. As mentioned earlier, the INTERPOL environmental crime programme functions exclusively on external grants and does not get support from member states. Same with CITES—though it receives contributions from member states and has a trust fund for the supporting staff, it also relies on external funding to enhance enforcement capacity development. Till 2010, a single enforcement officer mandated the matters of CITES enforcement and compliance process. But with more contributions coming in after the formation of ICCWC, CITES enhanced its staff capacity. None of the enforcement agencies would prefer a fund diversion, especially at a time when the need for their own activity is equally important to the need for information-sharing. Barnett and Coleman (2005) mention how, on one occasion, INTERPOL shifted its design to a set of strategies in order to pursue their goals in response to changing environmental pressures and constraints that potentially threaten their relevance and resource base.

Based on all the above reasons, it seems that so far the global-regional enforcement information-sharing process has been all about 'muddling through' rather than a scientific or systematic process of information verification, input and transfer. National authorities will hold on to information either due to data sovereignty or out of fear that they may lose a key negotiation position during multilateral meetings. In other words, when the decision in question is complex, for example, as in the case of ivory and rhino trade, information flow into an international agency will not take place smoothly as may be the case with the collection of weather data or other less complex environmental issues.

From all the above inferences, it can be concluded that, while actors within the wildlife policy subsystem perceive wildlife crime enforcement as a common good and have good intentions, they are unknowingly creating a mess. Within this mess, they all have a common objective of averting a problem—wildlife crime. But at the same time, their views differ, especially when it comes to the four distinct factors mentioned earlier (conflicting mandates, sovereignty, conflicting beliefs and the resource crunch). Technocratic scientists (including UNU researchers) often fail to realise this trend and generalise the problem as 'political'. Such naive assumptions can be seen in several scientific literatures which promote the use of open-data solutions in policymaking or conceptualise the relationship between science and politics as a linear process of knowledge transfer, research use or impact (Nutley et al., 2007). From our study, it is clear that such positivist approaches do not work in the case of enforcement information-sharing. The mess within the policy subsystem puts policymakers 'on top' (Hoppe et al., 2013) who then can call on the services of 'convenient' scientists who are 'on tap'. It is also naive to assume that all scientists are neutral, objective and independent experts, speaking 'truth to power'. Supporting the views of Jasanoff and Martello (2004) as well as Swedlow (2012), the neutrality of scientists can be questioned when they become aligned with different camps of beliefs, depending on their respective national or cultural perspectives or their countries' bargaining positions. In such a scenario, each stakeholder (including heads of UN agencies or enforcement bodies) will have their own different assumptions characterised by their own view or vision of addressing a problem.

Relating the above-mentioned factors with the case of adoption and implementation of WEMS, one can consider the case of transboundary enforcement-sharing as a *complex problem* where there is no true or false justification to why certain information systems are accepted at certain levels and objected at a different level.

Universal acceptance of a system remains an open-ended problem (see Chandran et al., 2015).

In other words, for WEMS to act as a 'boundary object', it is important for the system to place itself in a suitable configuration where it can bind the four factors in different scales (Global, Regional and Local).

6.4 In Search of a Solution…

After identifying the core reasons for conflicts in the adoption of WEMS, the UNU scientists and project managers of WEMS expanded to include a political scientist from the University of Twente and got together in identifying a suitable mechanism to carry out boundary work. The discussions aligned around terms like 'Persuasion', 'Rhetoric', 'Powering', 'Puzzling', 'Problem Structuring' and 'Problem Framing'. However, Persuasion and Rhetoric cannot bring long-standing solutions which WEMS initiative was seeking. Powering, Puzzling and Problem Framing, on the other hand, are part of the larger process of Problem Structuring. The team recognised that all methodological choices are problematic, unless mapping or abstraction of complex social processes is done based on real events related to the problem. Meaning, problem abstraction should be done by understanding the various phenomena that have taken place during problem structuring within the WEMS context since its conception.

Bringing in the concept of problem structuring (see Table 6.1) into the processes that has taken place in the development and implementation of WEMS, it can be seen that there was a constant process of renegotiations or '*structuring'*, as part of the larger political process where various stakeholders were considering how to react to objections from the opponents of such systems. With each objection, the WEMS system was undergoing transformations within the information design process based on the type of objections it received. The system design also changed because of conflicting scenarios with various actors.

The scenarios of structuring can be described as follows:

- Stage 1—UNU had to give up their initial preferred policy solution in addressing wildlife crime through a civil society network for *negotiating* a larger acceptance of WEMS within the policy subsystem. Though this changed the initial objective or purpose of WEMS, it helped to get the problem transformed into a moderately structured one where there was considerable normative consensus among CITES Parties and the CITES Secretariat.
- Stage 2—By requesting UNU to pilot test WEMS in Africa, WEMS got its first test implementation at the national and regional levels. Though there was no endorsement for WEMS, the test implementation can be considered as a stage of *applied research* where WEMS got an opportunity to showcase itself as a potential tool for regional enforcement monitoring.

- Stage 3—In the third stage, the focus of attention was drawn back in providing the onus of information collection to international enforcement agencies where CITES decided to use existing enforcement channels as a means of enforcement information-sharing. This led to the objection of WEMS at the global level. However, the regional implementation of WEMS in Africa continued. This shift brought back WEMS to Stage 1 where it had to *renegotiate* or restructure itself for an acceptance at a global level.

As mentioned earlier, the mismatch in the case of WEMS is that even during the time of value conflicts, international enforcement agencies, scientists and NGOs never denied the underlying advantage of the WEMS initiative and thought it had good potential in addressing the current challenges in enforcement information-sharing. Though global enforcement bodies were entrusted with the task of garnering information from national governments, such a mechanism was not working, as national agencies never provided complete information to a global body.

Hence, the only way to structure the problem in WEMS was to frame it as a tool that could bridge national-level information with regional and global levels.

6.5 Reframing WEMS as a Global Boundary Object

After identifying the core problems in the adoption of WEMS, UNU decided to host a forum to discuss the challenges and to identify a relative solution by using WEMS as a boundary object.

The first step was to invite the relevant global and national enforcement agencies, researchers working on WEMS, NGOs and government agencies implementing the system, to a common negotiation process. The initial idea was to have a workshop and prepare its agenda in consultation with all the relevant agencies, including CITES, INTERPOL, WCO and Lusaka Agreement. This was to ensure that the workshop followed certain criteria:

(a) Meaningful participation in agenda setting and knowledge production by stakeholders at the regional and global level.
(b) Governance arrangements that assure accountability.
(c) Production of collaborative products (boundary objects) such as a workshop recommendation which is both adaptable to different viewpoints and robust enough to maintain identity across them.

The overall purpose of the workshop was to identify an interface (see Hoppe, 2009; Jasanoff & Martello, 2004; Star & Griesemer, 1989) between organisations, disciplines, methods, culture and ideas, where knowledge is co-produced and a shared view is expressed between the actors, irrespective of the inclusive view of each entity. And, specifically to construct and manage effectively the interfaces among the four conflicting factors (listed in Sect. 6.3) that was constraining the adoption of WEMS.

It was for the first time in 10 years that UNU had initiated a workshop for addressing the challenges in enforcement information-sharing with such criteria and with all the relevant agencies being present, including CITES, INTERPOL, WCO and Lusaka Agreement. Until then, an *impermeable boundary* (Clark et al., 2011) had been blocking the interface between UNU and the CITES Secretariat, and no meaningful communication was taking place.

6.5.1 Participant Selection

In a situation of conflict between values, a process of mediation can only occur when there is a room for negotiation. Several literatures (Clark et al., 2011; Gieryn, 1983; Robinson & Wallington, 2012) on boundary work have used the term *demarcation* as a mechanism to build consensus. Demarcation is a process of protection from unwanted participants and interference, thereby prescribing proper ways of behaviour for participants and non-participants (Robinson & Wallington, 2012). Unlike large meetings such as Conference of Parties or Standing Committees, a demarcation process can best unfold in user-defined workshops where invited conflicting Parties are brought to the table for negotiation.

In this regard, the workshop participants were carefully selected. The criteria for inviting the participants were as follows:

(a) Should be familiar with wildlife law enforcement, has either engaged in high-level discussion or has contributed significantly to the process through scientific or political processes.
(b) The participant is familiar with wildlife enforcement monitoring tools and databases, including WEMS, I24/7, CEN database and EU-TWIX.
(c) They should have the skills and technique to bring out a technical or policy solution in addressing information-sharing.
(d) The participants come from international enforcement agencies (CITES, INTERPOL, WCO and UNODC), regional bodies (Lusaka Agreement) and Governments (Asia and Africa) who are the lead decision-makers on wildlife enforcement and compliance matters or on matters related to the CITES Convention itself.
(e) Representatives of the NGOs who were part of Asian Conservation Alliance and who were involved in the initial stages of the development of WEMS.

6.5.2 Problem Framing and Agenda Setting

Having confirmed the need for a workshop, the overall theme of the workshop was discussed. However, it was almost impossible to even imagine a bridging consensus on the beliefs, sovereignty, conflicting mandates and resource allocation, in one stretch. Especially in the context of a polarised policy subsystem, how can one event bring consensus on the beliefs of the various stakeholders?

The idea was then to avoid using the term 'conservation' from the concept of the workshop and to frame the problem as an 'enforcement', or security issue. By doing so, the attention on trade and conservation was diluted in the overall objective of the workshop.

After framing the problem, an advance communication was sent to CITES regarding the plan for the workshop. The UNU team of researchers and project managers then engaged with the ICCWC officer and sought advice on how to move ahead with the workshop. Similarly, discussions were held with other international enforcement agencies, including INTERPOL and World Customs Organisation, and based on their advice, the agenda for the workshop was finalised.

Once the agenda was set and the participants identified (UNU, 2014b) or demarcated, the next step was to build coordination between the agencies. The coordination process involved pre-negotiations and extensive communication, especially between 3 agencies—INTERPOL, WCO and CITES. Coordination also helped in defining proper ways for interaction between practices.

A few months prior to the workshop, United Nations had proclaimed the third of March as the World Wildlife Day (UN, 2013). This coincidence helped to bring a larger attention to the event. The dilemma was on whether the event should be called a workshop or a conference. Maintaining the concept of having a closed *working group* to discuss matters related to WEMS was important and the fear was that a larger conference would take attention away from the defined purpose. Keeping this in mind, the UNU team decided for a hybrid solution. The team formulated a three-day conference with the second and third days reserved exclusively for discussions on WEMS and enforcement information-sharing. The conference title was 'The Tokyo Conference on Combating Wildlife Crime' and was confirmed for March 3–5, 2014, in Tokyo.

The first day of the conference was planned to raise awareness on the importance of transboundary enforcement information. This included a panel discussion on:

(a) Evidence-based policymaking in combating wildlife crime.
(b) Role of the UN in combating wildlife crime.
(c) Learning lessons from Japan in combating wildlife crime.

The specific panels were also forums for extracting specific knowledge from scientists, the public and policymakers in the course of the discussions and deliberations on the second day.

The second-day working group sessions were planned as open only to the invited participants, who were then divided into three working groups to discuss:

- Secure information-sharing, data analysis and data mining.
- Science to policy linkages through WEMS initiative.
- Capacity building in developing countries to address information gaps.

A plenary session would then bring out the outcomes of the three working groups (linking the outcomes of each working group into one coherent document). The final output of the workshop was to be called the 'Tokyo Declaration on transboundary enforcement information-sharing on wildlife crime'.

The UNU team of researchers and project managers also planned to use the conference to negotiate with key actors sequentially and iteratively (rather than simultaneously) on identifying the key areas of concern in bringing consensus between the partners.

6.6 Tackling the Problem through Boundary Work

By January 2014, the momentum was picking up on addressing wildlife crime with several incidences of poaching taking place in Africa, and seizures happening in Asia. The British government hosted the London Conference on Illegal Wildlife Trade in February 12–13, 2014, where 46 countries agreed on a declaration containing a series of commitments, including addressing corruption and adopting legislation for tougher penalties against poachers (UK, 2014).

On March third 2014, the Tokyo conference on combating wildlife crime (UNU, 2014a) was held at UNU Headquarters, and it was chaired by Ms. Maria Mutagamba, Minister of Tourism, Wildlife and Heritage of Uganda (See Fig 6.2). The conference was presided over by the Secretary General of World Customs Organisation (WCO), senior enforcement officers from INTERPOL, WCO, the CITES Secretariat and Lusaka Agreement. Senior officials from the WEMS partnering countries also participated in the conference (Fig. 6.2).

The Tokyo Conference focused exclusively on enforcement information-sharing and bringing good governance. The participants attending the open-day session (day 1) unanimously recognised that there is a growing urgency to combat organised wildlife crime and that the need for a comprehensive information system is of utmost importance. They also acknowledged the importance of the existing information-sharing systems such as INTERPOL I24/7 and WCO CEN on a global scale, as well as more regional systems, such as WEMS in sub-Saharan Africa. The Tokyo Conference also acknowledged that WEMS has been a success and supported its further expansion to other interested States in the future (see UNU, 2014a & Recommendations of Tokyo Conference attached as Annex).

6.6.1 The Process of Boundary Work…

As Mcknight (2007) mentions, successful boundary objects are the products of multi-stakeholder development processes. In this regard, it was important to consider equal participation from all the invited members and to engage them in discussions relating to the use of the WEMS system.

During the working group meeting on the second day, officials from Lusaka Agreement presented scenarios on how WEMS was used by them and how specific

Fig. 6.2 The World Wildlife Day held on March 3 at UNU in Tokyo. Ms. Maria Mutagamba, Minister of Tourism, Wildlife and Heritage of Uganda, along with the Director of World Customs Organisation, Director of Division of Environmental law and Conventions, Director of UNU Campus computing, Professor of Policy and Knowledge from University of Twente and the Executive Director of an NGO (which was part of Asian Conservation Alliance), participating in the session titled *Evidence-based Policymaking in Addressing Wildlife Crime*

enforcement activities were carried out based on the information. During the time of the conference, 598 wildlife crime cases were recorded in the database from four African countries. Officials from World Customs Organisation and CITES agreed that the recordings indicate good progress on the initiative. The next steps were to identify how to link up the policy aspects of information-sharing between the regional database and the existing global databases (or linking up two social worlds). The working group would have been meaningless had they not discussed the issues of sovereignty, conflicting mandates of agencies and, most importantly, the issue of resource allocation.

6.6.2 Tackling Sovereignty Concerns

National enforcement agencies and officers from Wildlife divisions from Lusaka member state countries agreed that there was no problem in sending non-nominal information to global enforcement agencies and they can make immediate arrangements for the same. However, for providing nominal information to INTERPOL from the WEMS system they had reservations. The reason was that information to INTERPOL had to be shared by the National Central Bureau (NCB) of each country and the decision was made by NCB. Even if wildlife divisions from each country decides to send the details of a case to NCBs, the final call on information-sharing lies solely with the authority at NCB.

In policy terms, even if some functional flexibility can be obtained in sharing the data between all agencies, the lack of political flexibility causes a failure in the information flow. On the other hand, if both functional and political flexibility can be achieved through such workshops, then there is hope for successful adoption of WEMS by both regional and global bodies.

The way forward in solving this problem was to ask INTERPOL headquarters to convince the national authorities by highlighting the importance of getting nominal data to INTERPOL office. Finally, the workshop participants agreed that enforcement agencies and Lusaka Agreement should work closely to garner political will and flexibility to achieve this objective.

6.6.3 Addressing Conflicting Mandates

After achieving consensus on addressing the problems of sovereignty, the groups then discussed how to avoid duplication of work by different agencies. A need for more coordination was expressed by the representative from INTERPOL. If agencies are operating in isolation, then they may overlap with the task of another. CITES and World Customs Organisation agreed to strengthen the global coordination[4] system through the ICCWC forum which is a consortium of all global enforcement agencies working on wildlife crime.

At the national level, the concern was on being part of several systems, including INTERPOL I 24/7, CEN initiative and now WEMS. Most of these systems have similar mandates but not identical.

Two major decisions were made to address the problem:

(i) The WEMS system should not be operated in isolation and it should feed into the existing systems used by the global enforcement agencies.
(ii) Cases which are not relevant at the international level and without names of individuals need not be sent to INTERPOL.

As can be seen later, this decision was a key factor in the adoption of WEMS.

The final outcome of the meeting was to exclude NGOs in enforcement information collection, as transboundary information-sharing by NGOs could question the sovereignty of States. However, their role was defined well in the capacity development and awareness initiatives on strengthening enforcement information-sharing.

[4] INTERPOL and the United Nations Office on Drugs and Crime (UNODC) continue to establish inter-agency committees at the national level to enhance coordination and information-sharing among the relevant law enforcement authorities; for example, a National Environmental Security Task Force (NEST), Transnational Organized Crime Units (TOCUs), or other appropriate structures (extracted from the Recommendations that emerged from The Tokyo Conference on Combating Wildlife Crime).

6.6.4 *Agreement on Resource-Sharing*

After an agreement was reached on handling data sovereignty and conflicting mandates, the subject of resource allocation to regional enforcement bodies was an easy task. CITES mentioned that an endorsement letter from ICCWC can be provided to the Lusaka Agreement to raise funds provided they meet the requirement on sharing data. Other donors, including representatives from the Government of Japan, also promised to help and support national governments in building a datacentre in Africa to ensure sustainability of the WEMS initiative.

In conclusion, all the enforcement agencies agreed to the potential of WEMS but reminded that it should not be confined within regional agencies like the Lusaka Agreement. For it to gain its full potential, it should systematically feed to global systems like the WCO database and the INTERPOL system.

In the case of civil society, the most important issue was not just about data but more about the information on the hotspots of crime and how Parties were addressing compliance. They wanted to know how Parties were addressing the problems on poaching and illegal trade, and how much of it would be communicated to them. To bring in awareness, the Lusaka Agreement member states decided to make the seizure map produced by the WEMS system public so that civil society is aware about the work carried out by the national enforcement agencies.

Finally, a recommendation (owing to the fact that a declaration is usually signed by State heads) was drawn out of the findings from the working group which provided the first official endorsement of WEMS by the global enforcement bodies.

6.7 Conclusion

It is almost 20 years since the birth of WEMS. It has taken different courses during which it has constantly evolved and reshaped. One thing is quite clear—though the objection has been literally attributed to the WEMS system, the real objection is against a purely science-based solution which is not what CITES and other agencies are seeking. For them, it should be a mix of science and political convenience with the boundaries of engagement between various entities (NGOs, governments, international agencies and scientists) vaguely defined, allowing the actors to refute or support decisions when and where convenient. In this context, all the stakeholders involved in the functional and political context of an enforcement information system agreed that the best way to implement WEMS will be to interconnect with the existing global systems and not to work in isolation. The idea of development of a Digital Public Infrastructure for addressing Human-Nature Conflicts is also based on this premise.

6.7.1 Boundary Arrangement at the National Level

By implementing the project at the national and regional levels in Africa, WEMS had been able to bridge several silos of information into one common information platform (Fig. 6.3). This indicates that WEMS has been successful as a multi-stakeholder boundary object where it brought some form of congruence in the trust, beliefs and rules of various agencies at the national level. Thus, it could be embedded in an enterprise system and connected with other databases based on shared interests. For example, WEMS and ACLED databases complement well as it can map out the relationship between political conflicts and wildlife crime. Similarly, climate databases can indicate whether weather conditions or long-term climatic conditions have an impact on poaching incidences.

The *boundary arrangement* at the national level was achieved through demarcation and coordination, as listed below:

- National Governments will be the custodian of enforcement data—Demarcation.
- Data sharing is the right of the Government—Demarcation.

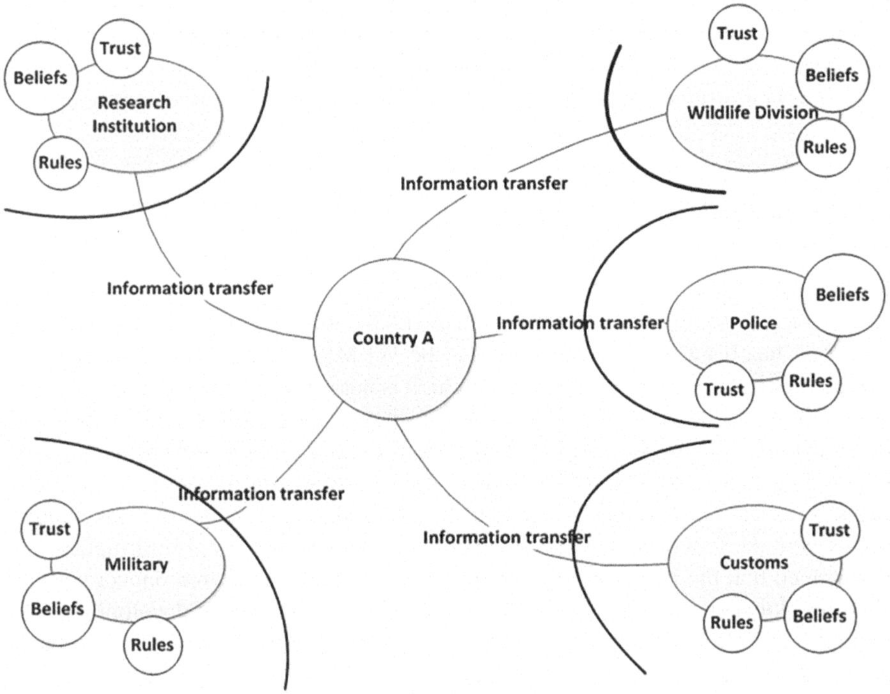

Fig. 6.3 WEMS as a multiple stakeholder boundary object Level – 1: National Context—WEMS as a national database

- Non-nominal information will be shared with the public—Coordination.
- Civil Society participation in capacity building—Coordination.

6.7.2 Boundary Arrangement at the Regional Level

One of the main reasons for the adoption of WEMS by Lusaka Agreement was due to a favourable policy subsystem at the regional level, where the Governing Council of Lusaka Agreement had already approved the decision to implement a suitable transboundary information system before WEMS was even presented at the Governing Council. This led to more coordination and less demarcation in defining *WEMS as a multiple stakeholder boundary object.*

The *boundary arrangement* at the regional level can be described as follows:

- National Governments decided to share wildlife crime data to LATF—Coordination.
- UNU and its associated partners carried out research on the data after consultation with the countries—Coordination.
- Research findings were provided back to countries for verification/approval—Demarcation.
- Civil Society was involved in capacity building efforts—Coordination.

By implementing WEMS in Africa, it was building coherence in translating the work of scientists to the policymakers belonging to two social worlds—national and regional levels (Fig. 6.4). The map (see Fig. 6.5) generated through the WEMS system, on the other hand, was the convergence tool that bridged the different stakeholders together.

The above map provides an overview of illegal wildlife trade seizure records from national enforcement agencies. The various coalitions within WEMS-Africa worked together to provide comprehensive information on illegal wildlife crime in sub-Saharan Africa. The red diamond labels highlight the seizure locations (data entered by national wildlife divisions) and the distributed dots show locations of ethnic conflict (data extracted from Harvard World Map).

6.7.3 Boundary Arrangement at the Global Level

In an international context, one of the challenges for global acceptance of WEMS was the difference in the regional and global policies. It was also observed that there were four factors, namely conflicting beliefs, conflicting mandates, data sovereignty and competition for financial resources, which were hindering the acceptance of WEMS. The Tokyo conference in a way was the boundary arrangement for the stakeholders at the national and international levels to come together and thereby

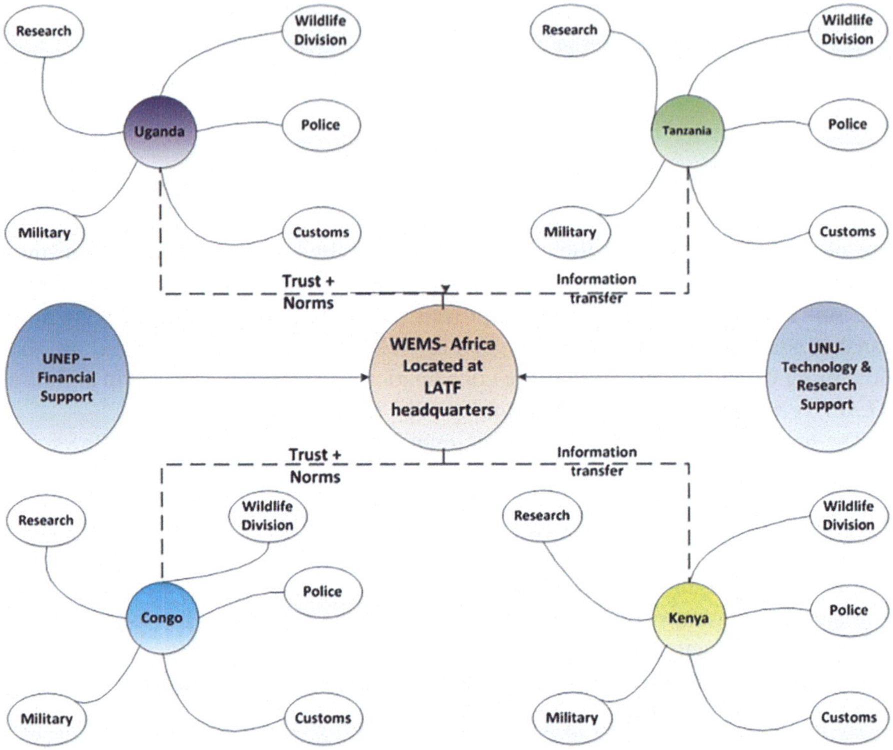

Fig. 6.4 WEMS as a multiple stakeholder boundary object Level-2: Regional Context—WEMS as a regional database

find a way forward in addressing their concerns. It was clear that global agencies, including CITES, INTERPOL and World Customs Organisation, never wanted WEMS to remain in isolation from the existing systems. At the same time, national governments were a bit reluctant to provide all the nominal data to international agencies. During the workshop on Transboundary Information-sharing (day 2), the agencies sat together on an intensive discussion, and finally the following recommendations were made on using WEMS and on information-sharing:

- Governments hold the right to disclose nominal information.
- Ecomessages developed through the WEMS system will be provided to NCBs which will then transfer it to INTERPOL.
- Reports of non-nominal information will be provided to World Customs Organisation in a quarterly manner.
- Support CITES management authorities in providing enforcement information to the CITES Secretariat.
- Map as an information-sharing platform for the public and civil society.

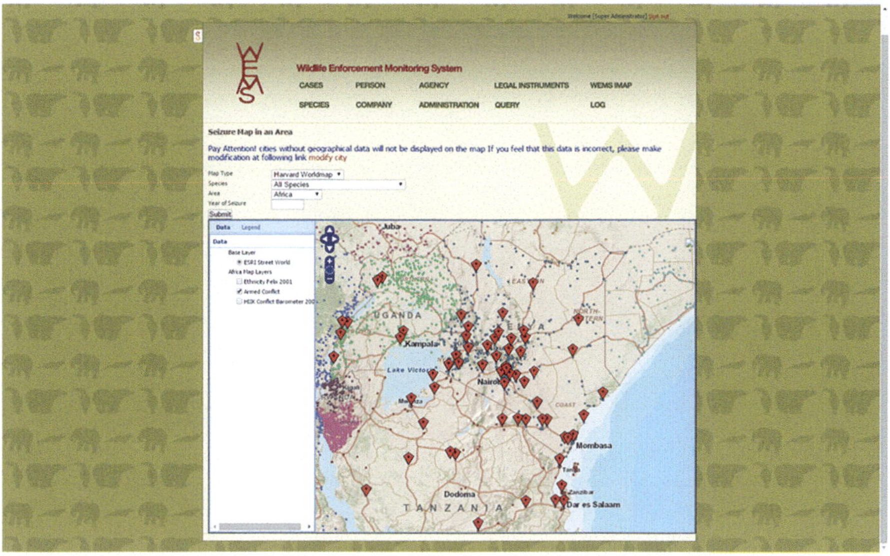

Fig. 6.5 Output Map (screen display) generated from WEMS-Africa

However, one should take into consideration that boundary objects are fragile and several externalities can bring changes to the existing arrangement in WEMS. The focus of the study has been until March 2014, and the externalities that can influence the wildlife policy subsystem in the future are however excluded from this book. A continuous and iterative process continued further which is explained in the next chapter.

Chapter 7
Moving Beyond Enforcement

7.1 Introduction

The preceding chapters have illustrated that, despite the proliferation of wildlife crime databases across the globe, significant shortcomings remain in rendering such data actionable for effective policy and enforcement interventions. The principal impediment lies not in the absence of technology, but rather in four persistent challenges: divergent policy beliefs, conflicting institutional mandates, issues surrounding data sovereignty and competition for limited financial resources—all of which have constrained the broader acceptance of the Wildlife Enforcement Monitoring System (WEMS). It has become increasingly evident that what is needed is robust *boundary work*—that is, mediating efforts to navigate the ideological and institutional divides that complicate transboundary information exchange. These chapters have further explored how regional and international policy divergences hinder progress in data interoperability and sharing.

Since 2014, WEMS has made limited headway in being recognised as a global model for monitoring wildlife law enforcement, and funding for the initiative has remained minimal. Many African governments, grappling with more immediate concerns such as public health crises, economic instability and the escalating impacts of climate change, have historically relegated wildlife crime to a lower rung of policy priority. This may also be attributed to a decline in poaching incidents within Lusaka Agreement member states (despite an increase in seizures), alongside a rise in human–wildlife conflict, which is becoming an increasingly pressing concern. Such a shift in focus has dampened both political will and financial support for combatting poaching and the illegal wildlife trade, including efforts to build comprehensive and systematic data collection frameworks.

Accordingly, the WEMS initiative has evolved, moving beyond a narrowly defined enforcement paradigm towards a more encompassing engagement with the broader spectrum of human–nature conflict. This transition reflects a growing

R. Chandran, *The Wildlife Enforcement Monitoring System*, Professional Practice in Governance and Public Organizations,
https://doi.org/10.1007/978-3-031-97084-9_7

awareness that conservation and wildlife protection cannot be pursued in isolation but must be integrated with socio-economic development, human rights considerations and environmental sustainability. This chapter examines the *fourth dimension* of policy evolution: the transition from monitoring wildlife crime to addressing human–nature conflict in its full complexity. This new dimension advocates a shift towards integrated information systems that do not prioritise a single issue, but instead focus on the interconnectedness of multiple drivers of conflict between humans and wildlife. The culmination of this approach is the proposition of a Digital Public Infrastructure (DPI)—an integrated system in which multiple databases converge to provide Digital Public goods that enhance our understanding and management of conflicts through insights into causal linkages and socio-economic transformations on a local, regional and global scale.

7.2 Human–Nature Conflicts: A Broader Context

Conflicts between human populations and the natural world are becoming ever more pronounced, as human expansion and economic activities encroach upon ecologically sensitive landscapes. These tensions manifest in various forms, exacerbated by accelerating global trends. Chief among these is climate-induced resource competition. As climate change intensifies, communities—particularly pastoralists and smallholder farmers—are compelled to move into new territories, often resulting in the encroachment of protected areas and wildlife habitats. This pattern is especially evident in regions like Africa, where climate change induced abnormal weather patterns are driving both humans and animals into increasingly constrained spaces, heightening tensions and the risk of violent encounters. In such scenarios, protected species may be killed, blurring the lines between subsistence hunting, illegal trade and conflict mitigation.

In many rural areas, subsistence poaching has become a survival strategy in response to economic hardship and food insecurity. Individuals, especially those lacking access to alternative livelihoods, may target high-risk species such as elephants and rhinoceroses—not out of malice, but as a last resort. Addressing this form of poaching presents a profound challenge, as it sits at the intersection of poverty, environmental degradation and inequality. Responses must therefore go beyond enforcement to incorporate poverty alleviation, community development and inclusive conservation models.

Human–wildlife conflict also arises from the spatial expansion of human settlements into areas traditionally inhabited by wildlife. As urbanisation encroaches on forests and grasslands, wildlife increasingly comes into contact with human populations, leading to the destruction of crops, predation of livestock and, in some cases, injury or loss of life. In numerous developing countries, such conflicts are intensifying as agricultural lands give way to urban sprawl, bringing large mammals—particularly elephants—into close proximity with human communities.

Further compounding the problem is the illegal exploitation of natural resources—such as unregulated logging, mining and wildlife trafficking—which continues to degrade ecosystems and endanger biodiversity, especially in regions with weak governance and enforcement capacity. Climate change serves both as a driver and amplifier of these pressures. Socio-economic factors such as poverty, unemployment and poor governance lie at the root of both illegal wildlife trade and agropastoral conflicts. These conflicts—typically arising from tensions between sedentary farmers and nomadic herders—have been on the rise since 2010 and are increasingly intertwined with broader patterns of insurgency and violence.

7.3 Data Gaps and the Need for Integrated Solutions

While human-nature conflicts are on the rise, and as previously discussed, governments have largely failed to prioritise the systematic collection of data relating to illegal wildlife trade, human–wildlife conflict and agropastoral tensions—due to both political reticence and technical limitations. Effectively addressing this complex nexus of human–nature conflict requires surmounting substantial obstacles to data availability, data collection, integration and accessibility. For instance, the lack of reliable, spatially explicit data on wildlife movement, poaching hotspots and patterns of human–wildlife interaction significantly hampers real-time response mechanisms—particularly those aimed at community protection. But most importantly, the culture of systematic data collection on nature based conflicts is missing in many developing countries. Further, investment in geospatial technology—essential for generating such insights—remains prohibitively expensive for many nations.

For data to be genuinely impactful, it must be transparent, accessible and interoperable across institutional, sectoral and national boundaries. However, disparate data standards and restrictive sharing policies often result in duplication of efforts and inefficiencies. Furthermore, ensuring responsible data governance—especially where local communities are concerned—requires stringent safeguards. Fear of persecution or retribution can prevent communities from sharing vital information unless their privacy and security are assured.

Beyond technical measures, there is a critical need to integrate climate data into systems designed to monitor human–nature conflicts. Climate change is altering wildlife migratory patterns and reshaping habitats, thereby increasing the likelihood of conflict. By incorporating climate variables, these systems can better anticipate future flashpoints and equip stakeholders with the tools to implement pre-emptive mitigation strategies.

In sum, the effective management of wildlife crime, human–wildlife conflict and agropastoral tensions—collectively referred to here as *human–nature conflicts*—is contingent upon access to timely, accurate and integrated data. Yet such data is often fragmented, inaccessible or simply non-existent. Without it, authorities are hamstrung in their ability to detect trends, allocate resources or design targeted interventions.

The WEMS research initiative has underscored the institutional difficulties faced by developing countries in managing and sustaining data infrastructures. During the Tokyo Conference on Combating Wildlife Crime, held at the United Nations University in 2014, African representatives expressed reservations about hosting enforcement data on private cloud platforms. These concerns centred on issues of data sovereignty and the potential commodification of sensitive information. Such concerns echo recent critiques, such as those put forward by Richard Pope in *Platformland (*Pope, 2024*),*[1] where he warns against adopting private-sector digital infrastructure models for public services—citing the fundamentally different logics of market mechanisms and public goods.

A notable recommendation emerging from the Tokyo conference was the establishment of a United Nations-led cloud infrastructure—one that would be trusted by member states and alleviate both financial and technical burdens. However, for any entity, including the UN, to assume such a role, a compelling rationale must exist—ideally one that directly supports the UN's Sustainable Development Goals. Moreover, for the UN, long-term operational funding and a dedicated support team would be essential to sustain such infrastructure.

A sustainable solution lies in broadening the pool of contributing data sources, especially by integrating existing databases from reputable institutions. As noted in earlier sections, any successful information system must function as a *boundary object*—capable of navigating and binding the diverse challenges of conflicting mandates, beliefs, sovereignty concerns and resource limitations across global, regional and local scales. By combining global and local datasets in an integrated framework, a more holistic, inclusive and resilient system can emerge—serving a broader range of stakeholders and enhancing the system's long-term viability. Nevertheless, the critical challenge remains: securing the technological, human and financial resources necessary to host such systems in a non-commercial, trusted environment.

7.4 United Nations Digital Public Infrastructure: A New Paradigm in Government Data Governance

The concept of a United Nations Digital Public Infrastructure (UN-DPI) presents a transformative opportunity for member states to address the multifaceted data governance challenges previously outlined. Envisioned as a centralised, integrated platform, the UN-DPI comprises a comprehensive stack of services, infrastructure and datasets designed to function as a decision-support tool for governments (see Fig. 7.1). This infrastructure would synthesise data from diverse sources—ranging from databases maintained by UN agencies, satellite imagery coordinated via the

[1] Pope (2024). *Platformland. An anatomy of next-generation public services*. London: London Publishing Partnership.

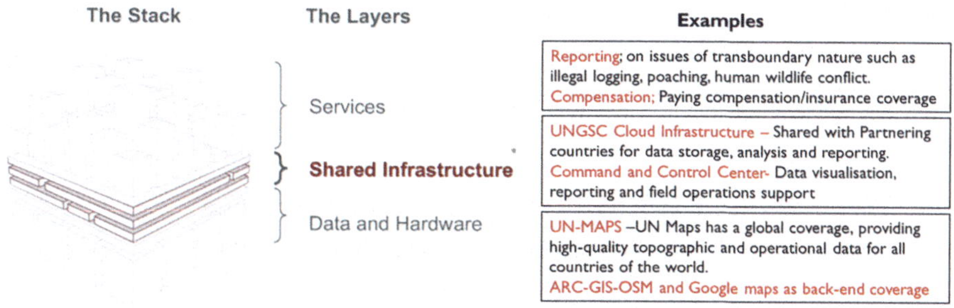

Fig. 7.1 The United Nations Digital Public Infrastructure (DPI) Stack and its layers (Figure modified from Eaves, D., Mazzucato, M., and Vasconcellos, B. (2024). Digital public infrastructure and public value: What is 'public' about DPI? UCL Institute for Innovation and Public Purpose, Working Paper Series (IIPP WP 2024–05). Available at: https://www.ucl.ac.uk/bartlett/public-purpose/wp2024-05)

United Nations Office for Outer Space Affairs (UNOOSA), ground patrol intelligence and the invaluable local knowledge contributed by national and sub-national authorities—thus enabling real-time access, analysis and responsive decision-making.

The strength of the UN-DPI lies not merely in its data collection capabilities but in its potential to enhance real-time coordination and response mechanisms. For instance, upon detection of an incident related to human-wildlife conflict or poaching, officials can use the platform to instantly notify the appropriate authorities, thereby enabling swift and coordinated action. Importantly, this shared infrastructure would foster cross-border collaboration and reporting—critical in the context of Africa's transboundary ecosystems—by facilitating interoperability and harmonised enforcement strategies. Furthermore, the system may be employed to forecast and mitigate agropastoral conflicts through climate-alert mechanisms that anticipate migratory shifts triggered by extreme weather events.

In addition, the UN-DPI would significantly bolster the capacity of local communities and enforcement agencies, many of which lack the requisite technical expertise or resources to fully utilise modern data tools. Through the provision of training programmes, analytical instruments and knowledge-exchange platforms, the DPI would empower stakeholders to engage more effectively with human–nature conflict resolution. Its multi-purpose nature enhances both its utility and its adoption.

However, recognising that information systems are not immune to political manipulation (as examined in Chap. 2), it is imperative that robust safeguards are embedded within the DPI framework. In this regard, the United Nations has already taken proactive steps, developing a universal DPI Safeguards Framework under the stewardship of the United Nations Office for Digital and Emerging Technologies

(ODET) and the United Nations Development Programme (UNDP).[2] This framework sets forth a comprehensive set of guidelines for the ethical design, deployment and management of digital public infrastructures.

7.5 Transition of WEMS from a Data Governance Instrument to a Digital Public Infrastructure

The Wildlife Enforcement Monitoring System (WEMS), originally designed to enhance data availability for law enforcement agencies, is now transitioning into a more comprehensive role within a broader Digital Public Infrastructure (DPI) framework for addressing human–nature conflicts. While WEMS focused on data utilisation for enforcement purposes, the DPI model emphasises the seamless flow of valuable information across multiple institutions. This approach brings in interoperability, establishes safeguards for data use and transfer and facilitates the interaction of disparate databases—thereby enabling richer analysis through the integration of varied information systems.

Insights drawn from the implementation of WEMS underscore a critical limitation: stand-alone systems focusing exclusively on environmental governance often fall short of capturing the complex, interconnected dynamics at play. For instance, the killing of an endangered species may result from illicit commercial motivations—constituting a clear case of poaching—or as a defensive response to threats posed by the animal to human life or livelihoods. Although both instances may be treated identically under poaching regulations, they stem from distinct causes and require differentiated responses. This reinforces the importance of integrated data systems that illuminate such nuances and enable more informed interventions.

Moreover, for any information system to be truly valuable, it must address human-centric concerns—an aspect that explains the continued appeal of community-based projects. Nevertheless, the process of expanding WEMS to incorporate additional datasets is not without challenges. It necessitates a broader bandwidth of collaboration among the custodians of various databases, all of whom must be willing to align their systems with the overarching DPI framework. In practical terms, WEMS must evolve into a modular component—capable of interfacing with other relevant systems, such as human–wildlife conflict databases and the Armed Conflict Location and Event Data (ACLED) platform.

The transition from WEMS to DPI followed a structured and participatory process undertaken by the team:

[2] UN Universal Safeguards for Inclusive Digital Public Infrastructure (2025). Retrieved April 14, 2025, from https://www.dpi-safeguards.org/

- As an initial step, the WEMS team convened a diverse group of stakeholders engaged in issues related to human–wildlife conflict, poaching, illegal wildlife trade and agropastoral tensions at the *Pan-African Conference on Strengthening Information Sharing Infrastructure and Governance Frameworks to Address Human-Nature Conflicts.*[3] This landmark event took place in Arusha from 17 to 18 June 2019 and was attended by representatives from African governments, UN agencies, intergovernmental and non-governmental organisations and policymakers and scientists.
- The primary objective of the conference was to examine the governance of nature-related conflicts through a unified platform for cooperation. It also highlighted the strategic role that national, regional and international entities could play in strengthening geospatial information-sharing infrastructure. The discussions emphasised the potential for such infrastructure to bridge the wildlife conservation and agropastoral policy subsystems across Africa.[4]
- Alongside the policy dialogue, technical experts showcased the possibilities presented by advances in spatial data infrastructure. These demonstrations illustrated how emerging technologies could inform and support governmental responses across the transboundary African landscape, all within the broader thematic context of managing human–nature conflict.

It was during this Pan-African conference that the notion of a regional Command-and-Control Centre (CCC) to address human–nature conflicts was first introduced. Since then, the concept has gained prominence and has been deliberated in several international forums, including the CITES Conference of the Parties (CoP) in Panama, the Tokyo International Conference on African Development (TICAD) and the United Nations Congress on Crime Prevention and Criminal Justice. Given that the Lusaka Agreement Task Force (LATF) is housed within the Kenya Wildlife Service premises, it has been proposed that the CCC could also serve as LATF's operational headquarters.

While governments reached a consensus on the necessity of establishing a CCC and a shared DPI tool, concerns were raised regarding the financial implications of maintaining large-scale ICT systems. Delegates also highlighted the challenge of limited network bandwidth across Africa, which constrains the transfer of high-volume geospatial data. During the conference, discussions were held on whether the United Nations could serve as the host platform for the DPI and in its

[3] Pan-African Conference on Strengthening Information Sharing Infrastructure and Governance Frameworks to Address Human-Nature Conflicts (2019). WEMS Initiative. https://wems-initiative.org/event/pan-african-conference-on-strengthening-information-sharing-infrastructure-and-governance-frameworks-to-address-human-nature-conflicts/

[4] Pan-African Conference on Strengthening Information Sharing Infrastructure and Governance Frameworks to Address Human-Nature Conflicts. (2019). Lusaka Agreement. https://lusakaagreement.org/pan-african-conference-on-strengthening-information-sharing-infrastructure-and-governance-frameworks-to-address-human-nature-conflicts/

secure cloud infrastructure. Given that the Pan-African conference had also received support from the Government of Japan, it was agreed to approach the Japanese government to assist in building the CCC and the development of the data infrastructure.

In pursuit of this objective, the Lusaka Agreement Task Force (LATF), in collaboration with the Government of the United Republic of Tanzania, the United Nations Development Programme (UNDP), the African Wildlife Foundation (AWF), the National Institute for Environmental Studies (NIES, Japan) and the Remote Sensing Technology Center (RESTEC) of Japan, convened a side event during the Seventh Tokyo International Conference on African Development (TICAD 7), held on 28 August in Yokohama, Japan. This event, followed by a press conference, was pivotal in spotlighting the importance of a spatial data infrastructure/DPI and a CCC mechanism in Africa.[5] The session was notably presided over by His Excellency Mr. Félix Tshisekedi, President of the Democratic Republic of the Congo. As the DRC is one of Africa's largest nations and custodian of vast natural resources, the President's presence lent considerable weight to the proceedings and garnered wider attention.[6]

Despite such high-level engagement, African governments were unable to reach an immediate decision on the location of the CCC. The Lusaka Agreement Task Force continued to advocate for Spatial Data Infrastructure (SDI)/DPI as a tool for law enforcement at various international forums, including the United Nations Crime Congress[7] and the Conference of the Parties to the Convention on International Trade in Endangered Species of Wild Fauna and Flora (CITES CoP19).[8] At each of these forums, remote sensing, SDI and geospatial technologies were consistently underscored as essential instruments for enhancing governance and enforcement capacities. Gradually, African governments began to acknowledge the value of these technologies in both conservation and enforcement contexts, marking a shift towards a more integrated approach to natural resource management.

In alignment with this evolving recognition, an extraordinary session of the 13th Governing Council of the Lusaka Agreement was held at the Emara Ole Sereni

---

[5] Extracted from the LATF press release at TICAD 7 Lusaka Agreement Task Force. (2019, August, 27). Transboundary Information Sharing and Governance in Africa: The need for the hour. Retrieved April 14, 2025, from lusakaagreement.org: https://lusakaagreement.org/wp-content/uploads/2019/08/LATFSideEventatTICAD7-1.pdf

[6] Strengthening Information Sharing Infrastructure and Governance Frameworks to address Human-Nature Conflicts, 28th August 2019—source: https://lusakaagreement.org/ticad7-side-event/

[7] Fourteenth United Nations Congress on Crime Prevention and Criminal Justice, Kyoto, Japan 7–12 March 2021, Source: https://wems-initiative.org/event/fourteenth-united-nations-congress-on-crime-prevention-and-criminal-justice-kyoto-japan-7-12-march-2021/

[8] Spatial data infrastructure for Addressing Data Gaps in Combating Illegal Wildlife Trade and Meeting Sustainable Development Goals in Africa "A Call to Action". https://wems-initiative.org/event/latf-cites-cop19/

Hotel in Nairobi, Kenya, in 2023. The Honourable Peninah Malonza, the then Cabinet Secretary for Tourism, Wildlife, and Heritage of the Republic of Kenya, assumed the presidency of the Lusaka Agreement.[9] The session was attended by member states including the Republic of the Congo, Kenya, Tanzania and Zambia, and concentrated on reinforcing the Lusaka Agreement and confronting transnational wildlife crime.

During this meeting, key discussions focused on the allocation of land for the Lusaka Agreement's regional CCC and on the hosting arrangements for the DPI platform to address human–nature conflicts. The Governing Council reviewed LATF's technical capacity to manage such infrastructure, with the conclusion that hosting services could be provided by either the United Nations or a qualified private entity—so long as strict adherence to the United Nations' DPI safeguards framework was ensured. Furthermore, the election of the Honourable Peninah Malonza reaffirmed Kenya's commitment to conservation. The Council undertook significant steps to strengthen the Lusaka Agreement and its Task Force, particularly in addressing illegal wildlife trafficking. The achievements of LATF were acknowledged, and member states were urged to renew their support for the organisation's mission.[10]

During the meeting, the Government of Kenya pledged land for the establishment of LATF's new headquarters and the Command-and-Control Centre in Nairobi, further solidifying its dedication to the Agreement. The Council also adopted a comprehensive Programme of Work and Budget for 2023–2024 and approved the creation of a Wildlife Enforcement Endowment Fund to sustain enforcement initiatives. Independent auditors were appointed to guarantee financial accountability. The meeting concluded with a strong commitment to ongoing collaboration and transparency in combatting the illegal wildlife trade.[11]

The Honourable Peninah Malonza played a key role in strengthening LATF during her tenure. Ms. Malonza proactively engaged with several African nations, encouraging them to join LATF and collaborate in transboundary conservation efforts. However, subsequent changes in Kenya's political landscape led to a reshuffling of the cabinet, resulting in her replacement—first by the Honourable Mr. Alfred Nganga Mutua and later by the Honourable Rebecca Miano. Although these changes temporarily slowed LATF's momentum, the Kenyan government's

[9] Kenya assumes Presidency as Parties to Lusaka Agreement Convene Extraordinary 13th Governing Council Meeting Source: https://lusakaagreement.org/kenya-assumes-presidency-as-parties-to-lusaka-agreement-convene-extraordinary-13th-governing-council-meeting/

[10] Kenya assumes Presidency as Parties to Lusaka Agreement Convene Extraordinary 13th Governing Council Meeting Source: https://lusakaagreement.org/kenya-assumes-presidency-as-parties-to-lusaka-agreement-convene-extraordinary-13th-governing-council-meeting/

[11] Kenya assumes Presidency as Parties to Lusaka Agreement Convene Extraordinary 13th GoverningCouncilMeetingSource:https://lusakaagreement.org/kenya-assumes-presidency-as-parties-to-lusaka-agreement-convene-extraordinary-13th-governing-council-meeting/

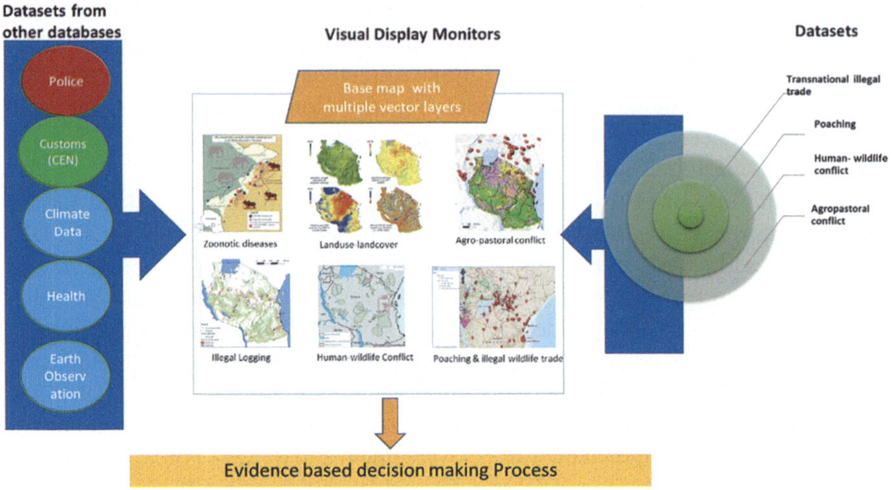

Fig. 7.2 Conceptual model of the Digital Public Infrastructure for human–nature conflicts

commitment to LATF remained steadfast. This was evidenced by the fulfilment of Kenya's pledge, made during the Governing Council session by Honourable Rebecca Miano, to allocate land for the new LATF headquarters and Command-and-Control Centre in 2024—a significant milestone in reinforcing Kenya's leadership in regional conservation. The brief yet impactful tenure of Peninah Malonza will also be remembered for catalysing this progress.

With the provision of land by the Government of Kenya, it was subsequently agreed that the DPI platform could be hosted at a suitable United Nations facility equipped with a robust ICT data centre. The idea of enlisting UN support for the SDI/DPI project was not new having first been proposed at the United Nations World Geospatial Information Congress, held in Hyderabad, India, in 2022.[12] LATF had previously encountered technical challenges in maintaining the system when hosted on its own premises in Nairobi.

Following the Kenyan government's formal land allocation for the LATF headquarters and Command-and-Control Centre, the Director of LATF, Mr. Edward Phiri, approached the United Nations Global Service Centre (UNGSC) to explore the possibility of hosting the DPI within the UNGSC's data infrastructure—thereby operationalising it as an UN-DPI. A detailed proposal for the UN-DPI framework for addressing human–nature conflicts was subsequently presented at the UNMAPS conference in Valencia, Spain, in 2024.[13] The presentation demonstrated how diverse

[12] Presentation of SDI/DPI at UNGGIM Source; https://ggim.un.org/2unwgic/documents/Remi%20Chandran.pdf

[13] The third annual UN Maps conference held in Valencia, Spain, from 20 to 24 May 2024. https://maps.un.org/un-maps-conferences/2024/presentations/un-maps-empowering-geospatial-digital-public-infrastructure

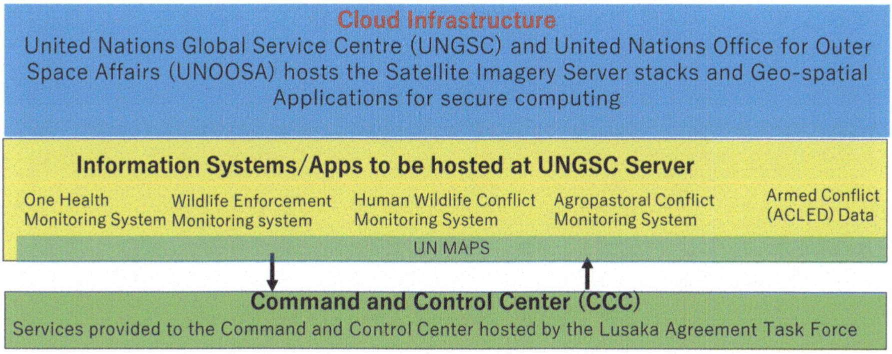

Fig. 7.3 Conceptual design of the Digital Public Infrastructure (DPI) (Stack for Human–Nature Conflicts, illustrating the data flow architecture from United Nations data repositories to the Command-and-Control Centre in Nairobi)

datasets on human–nature conflicts could be consolidated within a unified architecture, with artificial intelligence playing a central role in generating actionable insights tailored to user requirements (see Fig. 7.2). The architecture of the DPI Stack, its underlying cloud infrastructure, embedded datasets and dataflow mechanisms to the Command-and-Control Centre in Nairobi were all illustrated during the UN Maps Conference (see Fig. 7.3).

At the time of finalising this book, the Government of Kenya is also in dialogue with the Governments of India and Japan regarding potential financial support for the Digital Public Infrastructure (DPI) and the Command-and-Control Centre.

Epilogue: Towards a New Paradigm of Coexistence

In an age increasingly defined by the relentless march of environmental degradation and widening socio-economic divides, the resolution of wildlife crime and human–nature conflicts can no longer be approached solely through enforcement. Rather, it demands a holistic, data-driven strategy that places conservation within the larger framework of sustainable development. At the very core of this transformation lies the concept of Digital Public Infrastructure (DPI), a groundbreaking tool not merely for data acquisition, but for the reshaping of governance structures—ones that are inclusive, adaptive and forward-thinking.

DPI is more than a mere technological advancement; it is a means by which shared responsibility and collective intelligence can be fostered across sectors and borders. By weaving together geospatial data, real-time monitoring and community-driven insights, DPI allows decision-makers to respond with agility and precision to emerging threats, all the while ensuring that local knowledge and lived experiences are woven into the fabric of policy formulation. The convergence of technology and governance thus provides fertile ground for cross-border collaboration—an endeavour particularly critical in regions like sub-Saharan Africa, where ecological boundaries transcend the confines of national borders.

The triumph of DPI rests upon the bedrock of international cooperation. Initiatives such as the Tokyo International Conference on African Development (TICAD), the Pan-African Conference on Human–Nature Conflicts in Arusha (2019) and the 2021 UN Crime Congress have already laid the foundation for a shared vision of environmental security—one founded on principles of transparency and interoperability. These fora underscore the importance of aligning conservation efforts with the United Nations' Sustainable Development Goals (SDGs), particularly SDG 15 (Life on Land) and SDG 13 (Climate Action), viewing them not as regional concerns, but as a global imperative.

R. Chandran, *The Wildlife Enforcement Monitoring System*, Professional Practice in Governance and Public Organizations,
https://doi.org/10.1007/978-3-031-97084-9

From the lessons gleaned through the Wildlife Enforcement Monitoring System (WEMS) and its subsequent evolution into a DPI framework, six guiding principles emerge:

1. **Incentivise Data Contributions:** The creation of tangible incentives, such as the "One Dollar–One Record" model, motivates frontline actors to contribute to data ecosystems, thereby enhancing the depth and accuracy of monitoring efforts.
2. **Transform Governments into Data Scientists:** Continuous investment in professional development is essential. Empowering civil servants with analytical tools enables them to derive meaningful insights from data, shaping policies that reflect the realities of local contexts.
3. **Develop Transparent Data-Sharing Protocols:** User-friendly validation tools must accompany data-sharing frameworks to foster trust, clarify access parameters and mitigate institutional hesitation.
4. **Alleviate the Technical Burden on Governments:** The complexity of digital infrastructure should not fall solely on governments. Collaborative partnerships with technical providers can offer scalable, secure solutions.
5. **Maintain a Clear Thematic Focus:** The DPI should remain centred on the issue of human–nature conflicts. Avoiding thematic diffusion ensures that the platform remains practical, actionable and relevant to its core mission.

As this book draws to a close, I find myself reflecting on the realisation that the pursuit of a unified information-sharing framework such as WEMS was never intended to yield a perfect or permanent solution. Rather, it remains—first and foremost—an exercise in framing the possibilities of collaborative governance amidst entrenched institutional inertia and competing worldviews. What began as a vision to connect the disparate knowledge systems of governments, UN agencies, NGOs and scientists has blossomed into a broader discourse on the ethics, politics and pragmatics of transboundary data governance.

The journey has been anything but linear. It has required a continuous renegotiation of boundaries—be they intellectual, institutional or technological. At each juncture, WEMS met with resistance—not from a lack of purpose, but due to the deeply embedded beliefs and institutional loyalties of those it sought to serve. The challenge was not the absence of goodwill, but the multiplicity of worldviews and conflicting policy imperatives. In navigating these tensions, I have come to accept that while consensus may indeed be an illusion, congruence—partial, fragile, yet undeniably real—can be achieved through persistent framing, re-framing and the humility to know when to adapt.

If the transformation of WEMS into DPI has imparted one lesson, it is this: solutions to complex, cross-border environmental dilemmas cannot be conjured in policy vacuums or within the confines of institutional silos. They must be nurtured, through learning, compromise and the quiet resolve to persist when systems falter. My contribution, as both an insider and a reflective practitioner, has been to breathe life into a "non-living" artefact—endeavouring to align institutional interests towards a shared public good. Yet, like all policy artefacts, this work remains provisional scaffolding for what may come next.

The evolution of WEMS into the vision of Digital Public Infrastructure (DPI) marks a pivotal juncture: not a conclusion, but a branching pathway. DPI represents an emergent framework—responsive to the complexities of human–nature conflict and founded on the belief that information should serve not only power, but purpose. In this regard, it is only fitting to offer one final thought—a phrase that has emerged from years of dialogue, contestation and perseverance:

> Information systems will function only if they serve a human purpose.

In essence, the instinctive human drive to repair a faltering system is the reason why certain IT systems—such as those underpinning banking and food delivery—are so reliably maintained. In the realm of environmental monitoring systems, however, the true value of data lies not solely in its own metrics, but in its ability to positively affect human health and well-being. Beyond this, it is the power of data and analytics to inspire further funding, and to propel this mission forward, that grants it enduring value. Thus, for the success of DPI in addressing human–nature conflicts, it is imperative to think differently, act boldly and frame the issue anew.

For if there is one certainty in this uncertain endeavour, it is this: the work remains unfinished—and must, by its very nature, remain so, if it is to endure. This is not merely a conclusion, but an invitation. An invitation to governments, scientists, civil society and local communities to unite in the creation of a new paradigm—one where the divide between humanity and nature is no longer a site of conflict, but a frontier of cooperation. A future in which the health of our planet is not regarded as a peripheral concern, but as the bedrock upon which all human progress is built. In this vision, Digital Public Infrastructure for human–nature conflicts is not merely a tool—it is the architecture of hope.

Summary

In 2005, the United Nations University initiated the Wildlife Enforcement Monitoring System (WEMS) project in partnership with Asian Conservation Alliance. The idea of the project was to compile wildlife-seizure information through a crowd-sourced network and feed it into a geospatial infrastructure called WEMS with the support of the Asian Conservation Alliance, an NGO network. The move was criticised by CITES as the project accorded a significant role to conservation NGOs in Asia. Though the project was redesigned, its acceptance was hindered by several policy-related factors until WEMS finally got implemented in Africa.

In the quest for identifying the reasons for the objection of WEMS by CITES secretariat and its related implementation and adoption challenges, this study enquires into the political and policy problems that were hindering the adoption of WEMS as a geo-information tool for monitoring enforcement and compliance with the CITES Convention. The study traces the long trajectory of WEMS from its development to implementation in Africa. As the problems in WEMS were primarily policy related and not technological, the focus of the book has been largely on the policy aspect of its adoption.

Using policy theories including Advocacy Coalition Framework and Boundary Object Theory, this book first identifies the problems within the wildlife policy subsystem. Using Grid Group Cultural Theory and Q methodology, the study recognises four conflicting policy beliefs of coalitions involved in wildlife enforcement issues. Through the study, it was understood that one of the hurdles WEMS was facing was the polarisation of coalition beliefs within the wildlife policy subsystem. The other problems that stood in the way of WEMS implementation were conflicting mandates of enforcement agencies, sovereignty of States on national enforcement information and resource crunch triggering competition for financial resources between various coalitions working on wildlife enforcement issues.

Finally, in March 2014, through a practical application of the Boundary Object Theory, the study identifies a solution to the problem by bridging the stakeholders

R. Chandran, *The Wildlife Enforcement Monitoring System*, Professional Practice in Governance and Public Organizations,
https://doi.org/10.1007/978-3-031-97084-9

through a common framework, where WEMS acts as a 'boundary object', thereby serving the need of the major coalitions working on enforcement and compliance with CITES issues. Though the stakeholders agreed on the potential use of WEMS, the criteria for its final acceptance depended on how WEMS can serve as a feeder database to the existing global systems managed by WCO and INTERPOL.

In 2025, WEMS was redeveloped as a Digital Public Infrastructure model where it integrated multiple domains which worked on a common thematic area of human–nature conflicts.

This book concludes by observing that the pursuit of a unified information-sharing framework such as the Wildlife Enforcement Monitoring System (WEMS) was never intended to yield a perfect or permanent solution. Rather, it was—and remains—an exercise in framing the possibilities of collaborative governance in the face of entrenched institutional inertia and normative discord, where its forms and shapes will change with geopolitical influences and time.

Related Publications

Chandran, R., Alagesan, S. P., & de Vries, W. T. (2021). CITES enforcement information sharing—if you don't know where you've come from ... you don't know where you're going. Journal of International Wildlife Law & Policy, 24(3–4), 343–365. https://doi.org/10.1080/13880292.2022.2043410

Chandran, R., Hoppe, R., de Vries, W.T. and Georgiadou, P.Y. (2015). 'Conflicting policy beliefs and informational complexities in designing a transboundary enforcement monitoring system. In: Journal of cleaner production, 105(2015)', pp. 447-460

Chandran, R., Chong, N.G., Doll, C., Lee, Y.L., Mathai, V., Nguyen, K & Parayil, G. (2013). 'Bytes Beyond Borders: Strengthening Transboundary Information Sharing on Wildlife Crime through the Wildlife Enforcement Monitoring System (WEMS) Initiative', Policy Brief (pp. 8). Yokohama: United Nations University - Institute for Advanced Studies.

Chandran, R., Krishnan, P., & Nguyen, K. (2011). 'Wildlife Enforcement Monitoring System (WEMS): A solution to support compliance of Multilateral Environmental Agreements'. Government Information Quarterly, 28(2), 231-238.

Chandran, R. and Georgiadou, P.Y. (2010) 'Science, information technology and political convenience: the challenges in information sharing on illegal wildlife trade': abstract. In: Scaling and governance conference 2010: towards a new knowledge for scale sensitive governance of complex systems: conference program and book of abstracts. - Wageningen: Wageningen University and Research Centre (WUR), 2010. pp. 84

Remi Chandran, Adegboyega Ojo, Jean-Pierre Auffret, Ng Chong, Khoi Nguyen (2010)., Wildlife Enforcement Monitoring System (WEMS), Handbook- Version-1, United Nations University.

Chandran, R. (2010). Endangered Species Trade - Enforcement is Key, published in Our World 2.0. http://ourworld.unu.edu/en/endangered-species-trade-enforcement-is-key

R. Chandran, *The Wildlife Enforcement Monitoring System*, Professional Practice in Governance and Public Organizations,
https://doi.org/10.1007/978-3-031-97084-9

Media Quotes

1. Radio Netherlands (2011). Dutch technology combats trade in endangered species. http://www.rnw.nl/english/article/dutch-technology-combats-trade-endangered-species
2. Reuters (2007). New UN database to help combat wildlife crime. http://www.reuters.com/article/environmentNews/idUSL0476594320070604
3. Kyodo News (2006). UN University launches system to combat illegal wildlife trade. http://home.kyodo.co.jp/modules/fstStory/index.php?storyid=272256

Annex

Tokyo Conference on Combating Transboundary Wildlife Crime, United Nations University

Tokyo, Japan, 3-5 March 2014

Recommendations on Transboundary Enforcement Information Sharing on Wildlife Crime

The Tokyo Conference on Combating Transboundary Wildlife Crime was convened and hosted by the United Nations University in Tokyo, Japan, from 3 to 5 March 2014. Delegates from Congo, Japan, Kenya, Tanzania, Uganda and Zambia, representatives from United Nations University (UNU), United Nations Environment Programme (UNEP), Convention on International Trade of Endangered Species of Wild Fauna and Flora (CITES), World Customs Organization (WCO), International Criminal Police Organization (ICPO-INTERPOL), Lusaka Agreement Governing Council and Lusaka Agreement Task Force (LATF), national ministries of Uganda and of Japan, Kenya Wildlife Service (KWS), Uganda Wildlife Authority (UWA), Wildlife Division (WD)—Tanzania, Tanzania National Parks (TANAPA), Ngorongoro Conservation Area Authority (NCAA), Congo National Bureau to the Lusaka Agreement (BNAL), Zambia Wildlife Authority (ZAWA), Japan International Cooperation Agency (JICA), Remote Sensing Technology Center (RESTEC) of Japan, International Fund for Animal Welfare (IFAW), Japan Tiger and Elephant Fund (JTEF), TRAFFIC Japan and other relevant experts and organisations met with the objective to identify and promote good practices to enhance the fight against wildlife crime through effective sharing of transboundary enforcement information on wildlife crime.

Considering that:

- Wildlife crime continues to be a growing problem worldwide, with strong evidence of increased involvement of organised crime groups that operate through well-developed criminal networks.
- As a result of the involvement of organised crime groups, law enforcement authorities around the world are facing increasingly difficult and complex situations in the investigation of wildlife crime.

- Wildlife crime is posing a significant challenge to the conservation and protection of wild fauna and flora, and that it has become a serious threat to the economy, natural resources and cultural heritage of many countries, and deserves to be treated as a serious crime.
- There is a lack of reliable data on the nature and scale of illegal wildlife trade.
- There is a need to harmonise data standards to facilitate increased gathering and sharing of data and information among the national, regional and international levels.
- The United Nations University, located in Tokyo, Japan, developed WEMS as a tool to provide a common platform through which data on wildlife crime can be gathered, analysed and shared to assist governments by enhancing their enforcement efforts to combat wildlife crime.
- The Lusaka Agreement Task Force (LATF) has successfully piloted the Wildlife Enforcement Monitoring System (WEMS) in four Lusaka Agreement member states, and that the system will be expanded to all Lusaka Agreement members states as well as to other interested States.
- Systems such as WEMS should adhere to information exchange protocols and standards used or recommended by INTERPOL and the World Customs Organization (WCO), and that they serve to complement the existing databases maintained by these organisations.
- Sustainable funding is critical to effective transboundary enforcement information-sharing on wildlife crime.

We, the participants in the Tokyo Conference on Combating Wildlife Crime, recommend that:

Secure Information-Sharing

- INTERPOL and the United Nations Office on Drugs and Crime (UNODC) continue to establish inter-agency committees at the national level to enhance coordination and information-sharing among relevant law enforcement authorities, for example a National Environmental Security Task Force (NEST), Transnational Organised Crime Units (TOCUs) or other appropriate structures.
- Law enforcement agencies continue to use secure global communication systems and databases such as the INTERPOL-I-24/7 and the WCO Customs Enforcement Network (WCO CEN) to gather, analyse and share information on wildlife crime at international level, and as appropriate, consider the use of other information systems such as the Wildlife Enforcement Monitoring System (WEMS), to gather, analyse and share information on wildlife crime, as appropriate, at national or regional level.
- LATF, its partners and Lusaka Agreement member states continue to implement WEMS through the WEMS Africa initiative and that they update CITES, INTERPOL, WCO and UNEP on their progress.
- Law enforcement agencies use secure and international protocols and standards to facilitate interoperability of data among different systems.

Data Analysis

- Agencies responsible for wildlife law enforcement, and as appropriate, other role players, promote the submission of data to appropriate databases for analysis, to improve the current understanding of the scale and nature of wildlife crime, to enable the formulation of appropriate law enforcement responses and to promote the allocation of adequate resources to suppress it.
- Agencies responsible for wildlife law enforcement draw upon existing tools (e.g. INTERPOL Notices, WCO CENcomm and WCO ENVIRONET) to share information and intelligence on, for example, wanted criminals, modus operandi and illegal trade routes.
- Agencies responsible for wildlife law enforcement, as appropriate, continue to engage with universities, academia, non-governmental organisations (NGOs) and the private sector to examine the role of new technologies to enhance effective mining of big data from all available sources to combat wildlife crime.

Science–Policy Interface

- Agencies responsible for wildlife law enforcement to increasingly explore opportunities to engage with NGOs, local and rural communities, academic institutions and the private sector, recognising that they can play more important roles in buttressing efforts and actions against illegal wildlife trade.
- Agencies responsible for wildlife law enforcement, other experts and organisations increasingly draw upon databases such as the Monitoring of Illegal killing of Elephants (MIKE) and the Elephant Trade Information System (ETIS) to promote scientific research and analysis and to encourage the highest political support to combat illegal wildlife trade.
- Agencies responsible for wildlife law enforcement and conservation establish effective working relationships with leading national and international research institutions to support long-term policymaking.

Capacity Development

- Countries, if appropriate, make use of technical resources such as the ICCWC Wildlife and Forest Crime Analytic Toolkit to conduct an assessment of national capacity to respond to wildlife crime and to act upon its outcomes.
- The International Consortium on Combating Wildlife Crime (ICCWC) and, as appropriate, other experts and organisations increasingly conduct Training of Trainers initiatives on data and information management with reference to the

use of open and distance education resources such as those provided by the CITES Virtual College.
- ICCWC and, as appropriate, other relevant experts and organisations support regional and national wildlife enforcement agencies to acquire appropriate Information and Communication Technology infrastructure to include standardised hardware and software as well as develop training programmes on the use of new information and communication technologies and on analysis and modelling of data.

Funding for information management in wildlife law enforcement

The international community and governments provide financial support for the implementation of the above recommendations, as may be required.

References

Akrich, M., Callon, M., Latour, B., & Monaghan, A. (2002). The key to success in innovation part I: The art of Interessement. *International Journal of Innovation Management, 06*(02), 187–206. https://doi.org/10.1142/S1363919602000550

Andresen, S., Walløe, L., & Rosendal, G. (2005). 'The precautionary principle: Knowledge counts but power decides. In R. Cooney & B. Dickson (Eds.), *Biodiversity and the precautionary principle: Risk and uncertainty in conservation and sustainable use* (pp. 39–55). Earthscan.

Arnold, F. E. (2003). Native forest policy in Chile: Understanding sectoral process dynamics in a country with an emerging economy. *International Forestry Review, 5*(4), 317–328. https://doi.org/10.1505/ifor.5.4.317.22662

Avgerou, C. (2008). Information systems in developing countries: A critical research review. *Journal of Information Technology, 23*, 133–146. https://doi.org/10.1057/palgrave.jit.2000136

Bale, L. S. (2003). *Gregory Bateson's theory of mind: Practical applications to pedagogy.* Accessed Apr 11, 2016, from http://www.narberthpa.com/Bale/lsbale_dop/gbtom_patp.pdf

Barnett, M., & Coleman, L. (2005). Designing police: INTERPOL and the study of change in international organizations. *International Studies Quarterly, 49*(4), 593–620. https://doi.org/10.1111/j.1468-2478.2005.00380.x

Barnett, M. N., & Finnemore, M. (1999). The politics, power, and pathologies of international organizations. *International Organization, 53*(04), 699–732. https://doi.org/10.1162/002081899551048

Bechky, B. A. (2003). Sharing meaning across occupational communities: The transformation of understanding on a production floor. *Organization Science, 14*(3), 312–330. https://doi.org/10.2307/4135139

Bohlen, C., & Sandalow, D. (2002). Protecting mahogany: Bush policy sells Amazon treasure down the river. *New York Times*. Accessed Jun 30, 2014, from http://www.nytimes.com/2002/11/13/opinion/13iht-edcurtis_ed3_.html

Bonner, J. (1994). Battle for Brazilian mahogany. *New Scientist, 144*(1948), 16–17.

Bowker, G. C., & Star, S. L. (1999). *Sorting things out: Classification and its consequences (inside technology)*. MIT Press.

Brockington, D., & Scholfield, K. (2010). The conservationist mode of production and conservation NGOs in sub-Saharan Africa. *Antipode, 42*(3), 551–575. https://doi.org/10.1111/j.1467-8330.2010.00763.x

Brown, S. R. (1996). Q methodology and qualitative research. *Qualitative Health Research, 6*(4), 561–567. https://doi.org/10.1177/104973239600600408

R. Chandran, *The Wildlife Enforcement Monitoring System*, Professional Practice in Governance and Public Organizations,
https://doi.org/10.1007/978-3-031-97084-9

Cairns, R. C. (2012). Understanding science in conservation: A Q method approach on the Galápagos Islands. *Conservation and Society, 10*(3), 217–231. https://doi.org/10.4103/0972-4923.101835

Callon, M., & Law, J. (1982). On interests and their transformation: Enrolment and counter-enrolment. *Social Studies of Science, 12*(4), 615–625. https://doi.org/10.1177/030631282012004006

Carlton, L. J. (2007). *Map making and map use in a multi-actor context: Spatial visualizations and frame conflict in regional policy making in The Netherlands* (PhD dissertation), Printed by JB&A grafische communicatie, Delft.

Cattell, R. B. (1966). The scree test for the number of factors. *Multivariate Behavioural Research, 1*(2), 245–276. https://doi.org/10.1207/s15327906mbr0102_10

Chandran (2024). *UN Maps Empowering Geospatial Digital Public Infrastructure,* presented at the 3rd annual UN Maps conference 2024 in Valencia, Spain. https://maps.un.org/un-maps-conferences/2024/presentations/un-maps-empowering-geospatial-digital-public-infrastructure

Chandran, R., Alagesan, S. P., & de Vries, W. T. (2021). CITES enforcement information sharing—If you don't know where you've come from … you don't know where you're going. *Journal of International Wildlife Law & Policy, 24*(3–4), 343–365. https://doi.org/10.1080/13880292.2022.2043410

Chandran, R., Chong, N. G., Doll, C., Lee, Y. L., Mathai, V., Nguyen, K., & Parayil, G. (2013). Bytes Beyond Borders: Strengthening transboundary information sharing on wildlife crime through the wildlife enforcement monitoring system (WEMS) initiative. In *Policy brief* (p. 8). United Nations University–Institute for Advanced Studies.

Chandran, R., Hoppe, R., Vries, W. T., & Georgiadou, Y. (2015). Conflicting policy beliefs and informational complexities in designing a transboundary enforcement monitoring system. *Journal of Cleaner Production, 105*, 447–460. https://doi.org/10.1016/j.jclepro.2014.12.068

Chandran, R., Krishnan, P., & Nguyen, K. (2011). Wildlife enforcement monitoring system (WEMS): A solution to support compliance of multilateral environmental agreements. *Government Information Quarterly, 28*(2), 231–238. https://doi.org/10.1016/j.giq.2010.09.002

Chang, Y.-T., Chu, Y.-H., & Huang, M.-H. (2006). The uneven growth of democratic legitimacy in East Asia. *International Journal of Public Opinion Research, 18*(2), 246–255. https://doi.org/10.1093/ijpor/edl010

Child, B. (1996). The practice and principles of community-based wildlife management in Zimbabwe: The CAMPFIRE programme. *Biodiversity and Conservation, 5*(3), 369–398.

Chisholm, D. (1995). Problem solving and institutional design. *Journal of Public Administration Research and Theory, 5*(4), 451–492.

Christy, B. (2012). 'Blood Ivory'. *National Geographic*. Accessed Apr 20, 2025, from https://www.nationalgeographic.com/magazine/article/blood-ivory

CITES. (1979). *Speech by the executive director of the United Nations environment Programme, Dr. Mostafa K. Tolba*. Second meeting of the Conference of the parties. Accessed Apr 20, 2025, from https://cites.org/sites/default/files/eng/cop/04/E04-Opening-speeches.pdf

CITES. (1987). *Interpretation and implementation of the convention.* Doc (6.9), 541-565. Accessed Apr 20, 2025, from https://cites.org/sites/default/files/eng/cop/06/doc/E06-19.pdf

CITES. (1989). *Interpretation and implementation of the convention -review of alleged infractions*. Seventh meeting of the Conference of the parties, 602-651. Accessed Apr 20, 2025, from https://cites.org/sites/default/files/eng/cop/07/doc/E07-20.pdf

CITES. (1992). *Interpretation and implementation of the convention. Review of alleged infractions and other problems of enforcement of the convention.* Accessed Apr 20, 2025, from https://cites.org/sites/default/files/eng/cop/08/doc/E-19.pdf

CITES. (1994a). *Interpretation and implementation of the convention.* COP 9. Accessed Apr 20, 2025, from https://cites.org/sites/default/files/eng/cop/09/doc/E9-Doc-23-25.pdf

CITES. (1994b). *Thirty-first meeting of the standing committee. Summary report-SC31.* Accessed Apr 20, 2025, from https://cites.org/sites/default/files/eng/com/sc/31/E31-SumRep.pdf

CITES. (1995). *Notification to parties. ICPO-Interpol, the ECO-MESSAGE.* Accessed Apr 20, 2025, from https://cites.org/sites/default/files/eng/notif/1995/851.txt

CITES. (1997a). *Interpretation and implementation of the convention–enforcement matters*. Report by the secretariat on implementation of resolution Conf. 9.8, COP 10. Accessed Apr 20, 2025, from https://cites.org/sites/default/files/eng/cop/10/doc/E10-28to30.pdf

CITES. (1997b). *Trade in African elephant specimens -conditions for resumption of trade in ivory and the disposal of ivory stockpiles from African elephant range states*. Interpretation and implementation of the convention. Accessed Apr 20, 2025, from https://cites.org/sites/default/files/eng/cop/10/doc/E10-44to46.pdf

CITES. (2000a). *Conservation of and trade in elephants—monitoring of illegal trade and illegal killing*. Interpretation and implementation of the convention—doc. 11.31.2. Accessed Apr 20, 2025, from https://cites.org/sites/default/files/eng/cop/11/doc/31_2.pdf

CITES. (2000b). *Enforcement—review of alleged infractions and other problems of implementation of the convention*. Interpretation and implementation of the convention. https://cites.org/sites/default/files/eng/cop/11/doc/20_01.pdf

CITES. (2000c). *Summary report of the committee i meeting*. Accessed Apr 20, 2025, from https://cites.org/sites/default/files/eng/cop/11/other/Com_I.pdf

CITES. (2002a). *General compliance issues—enforcement matters. Interpretation and implementation of the convention (COP12 doc. 27)*. Accessed Apr 20, 2025, from https://cites.org/sites/default/files/eng/cop/12/doc/E12-27.pdf

CITES. (2002b). *Opening remarks by Mr Klaus Töpfer, Executive Director of UNEP*. CITES COP 12. Accessed Apr 20, 2025, from https://cites.org/eng/cop/12/topfer_open.shtml

CITES. (2003). *The big leaf mahogany and CITES appendix III. CITES world–official newsletter of CITES*. Accessed Apr 20, 2025, from https://cites.org/sites/default/files/eng/news/world/11.pdf

CITES. (2004). *General compliance issues–Enforcement Matters Bangkok: CITES*. Accessed Apr 20, 2025, from https://cites.org/sites/default/files/eng/cop/13/doc/E13-23.pdf

CITES. (2005). *Control of trade in African elephant ivory*. In CITES (Ed.), Fifty-third meeting of the standing committee–interpretation and implementation of the convention–African elephant. : CITES. Accessed Apr 20, 2025, from https://cites.org/sites/default/files/eng/com/sc/53/E53-20-1.pdf

CITES. (2007a). *CITES Compliance Procedures, Conf. 14.3*. Accessed Apr 20, 2025, from https://cites.org/sites/default/files/documents/COP/19/resolution/E-Res-14-03-R19.pdf

CITES. (2007b). *Interpretation and Implementation of the Convention—Compliance and enforcement issues*. Accessed Apr 20, 2025, from https://cites.org/sites/default/files/eng/cop/14/doc/E14-25.pdf

CITES. (2007c). *Summary Record (SC55) Fifty-fifth meeting of the Standing Committee*. Accessed Apr 20, 2025, from https://cites.org/sites/default/files/eng/com/sc/55/E55-SumRec.pdf

CITES. (2007d). *Summary record of the 14th session of committee II. CITES*. Accessed Apr 20, 2025, from https://cites.org/sites/default/files/eng/cop/14/rep/E14-Com-II-Rep-14.pdf

CITES. (2009). *CITES SC 58 doc. 23 (addendum): Interpretation and implementation of the convention, compliance and enforcement, report of the CITES enforcement expert group*. Accessed Apr 20, 2025, from https://cites.org/sites/default/files/eng/com/sc/58/E58-23Add.pdf

CITES. (2010a). *Interpretation and implementation of the convention; compliance and enforcement; enforcement matters*. CITES. Accessed Apr 20, 2025, from https://cites.org/sites/default/files/eng/cop/15/doc/E15-24.pdf

CITES. (2010b). *Powerful alliance to fight wildlife crime comes into effect*. Accessed Apr 20, 2025, from https://cites.org/eng/news/pr/2010/20101123_ICCWC.php

CITES. (2011a). *Gathering and analysis of data on illegal trade. Sixty-first meeting of the standing committee*. Accessed Apr 20, 2025, from https://cites.org/sites/default/files/eng/com/sc/61/E61-31.pdf

CITES. (2011b). *Gathering and analysis of data on illegal trade–Results of GLOBAL-TWIX questionnaire Geneva: CITES*. Accessed Apr 20, 2025, from https://cites.org/sites/default/files/common/com/sc/61/E61-31-A.pdf

CITES. (2011c). *Interpretation and implementation of the convention, compliance and enforcement.* Geneva: CITES. Accessed Apr 20, 2025, from https://cites.org/sites/default/files/eng/com/sc/61/E61-30.pdf

CITES. (2012a). *National laws for implementation of the convention. Notification to the parties.* Accessed Apr 20, 2025, from https://cites.org/sites/default/files/eng/notif/2012/E036.pdf

CITES. (2012b). *Sixty-second meeting of the Standing Committee; Interpretation and implementation of the Convention, Compliance and enforcement, National reports.* Accessed Apr 20, 2025, from https://cites.org/sites/default/files/eng/com/sc/62/E62-24-01.pdf

CITES. (2012c). *Submission of national reports.* Notification to the parties. Accessed Apr 20, 2025, from http://www.cites.org/eng/notif/2012/E016.pdf

CITES. (2012d). *What is CITES?* Accessed Apr 20, 2025, from https://cites.org/eng/disc/what.php

CITES. (2013a). *The elephant trade information system (ETIS).* What is ETIS? Accessed Apr 20, 2025, from https://cites.org/eng/prog/etis

CITES. (2013b). *How is CITES financed?* Accessed Apr 20, 2025, from http://www.cites.org/eng/disc/fund.php

CITES. (2013c). *International consortium on combating wildlife crime.* Accessed Apr 20, 2025, from https://cites.org/sites/default/files/eng/cop/16/doc/E-CoP16-15.pdf

CITES. (2013d). *National reports.* Interpretation and implementation of the convention–compliance and enforcement. Accessed Apr 20, 2025, from https://cites.org/sites/default/files/eng/cop/16/doc/E-CoP16-30.pdf

CITES. (2013e). *Secretary–general's certificate of commendation. Notification to the parties.* Accessed Apr 20, 2025, from https://cites.org/sites/default/files/eng/notif/2013/E--Notif-2013-014.pdf

CITES. (2014). *Big-leaf Mahogany.* Accessed Apr 20, 2025, from https://cites.org/eng/prog/mwg.php

CITES. (2025). *Appendices.* Accessed Apr 14, 2025, from https://cites.org/eng/app/appendices.php

Clark, W. C., Tomich, T. P., van Noordwijk, M., Guston, D., Catacutan, D., Dickson, N. M., & McNie, E. (2011). Boundary work for sustainable development: Natural resource management at the consultative group on international agricultural research (CGIAR). *National Academy of Sciences of the United States of America, 113*, 4615. https://doi.org/10.1073/pnas.0900231108

Clarke, A. H. (2002). Understanding sustainable development in the context of other emergent environmental perspectives. *Policy Sciences, 35*(1), 69–90. https://doi.org/10.1023/a:1016067819764

Collins, H. M., & Evans, R. (2002). The third wave of science studies. *Social Studies of Science, 32*(2), 235–296. https://doi.org/10.1177/0306312702032002003

Danaher, M. (Ed.). (2008). *Environmental politics in Japan: The case of wildlife preservation.* VDM Verlag.

Dawes, S. S., Cresswell, A. M., & Pardo, T. A. (2009). From "need to know" to "need to share": Tangled problems, information boundaries, and the building of public sector knowledge networks. *Public Administration Review, 69*(3), 392–402. https://doi.org/10.1111/j.1540-6210.2009.01987_1.x

Douglas, M. (1978). Cultural bias. In M. Douglas (Ed.), *In the active voice* (Vol. 35, pp. 183–254). Routledge and Kegan Paul.

Douglas, M., & Wildavsky, A. B. (1983). *Risk and culture: An essay on the selection of technical and environmental dangers.* University of California Press.

Dreher, A., & Fuchs, A. (2015). Rogue aid? An empirical analysis of China's aid allocation. *Canadian Journal of Economics/Revue canadienne d'économique, 48*(3), 988–1023.

Dryzek, J. S. (2001). Legitimacy and economy in deliberative democracy. *Political Theory, 29*(5), 651–669. https://doi.org/10.2307/3072533

DSWF. (2014). '*Our projects–Lusaka agreement task force (LATF)*'. Accessed May 30, 2025, from http://www.davidshepherd.org/project/000016/crime/

Duffy, R. (2000). *Killing for conservation: Wildlife policy in Zimbabwe.* International African Institute in association with J. Currey.

Duffy, R. (2013). Global environmental governance and north–south dynamics: The case of the CITES. *Environment and Planning C: Government and Policy, 31*(2), 222–239.

Dühr, S. (2007). *The visual language of spatial planning: Exploring cartographic representations for spatial planning in Europe*. The RTPI Library Series. Routledge.

Dunn, W. N., Mitroff, I. I., & Deutsch, S. J. (1981). The obsolescence of evaluation research. *Evaluation and Program Planning, 4*(3–4), 207–218. https://doi.org/10.1016/0149-7189(81)90022-7

Eckholm, E. (1987, July 28). 'Despite setbacks, world control of trade in wildlife expands influence'. *New York Times*. Accessed Mar 23, 2012, from http://www.nytimes.com/1987/07/28/science/despite-setbacks-world-control-of-trade-in-wildlife-expands-influence.html?pagewanted=all

Edge, J. (2015). *Elephant Charities, the Good, Bad and Ugly–What are NGOs doing to stop the slaughter of African Elephants?–See more at*: http://magazine.africageographic.com/weekly/issue-42/elephant-charities-the-good-the-bad-the-ugly/#sthash.D9Jdtw1j.dpuf. Africa Geographic Magazine. 47.

Elliott, L. (2007). Transnational environmental crime in the Asia Pacific: An 'un(der)securitized' security problem? *The Pacific Review, 20*(4), 499–522.

Emad, G., & Roth, W. (2009). Policy as boundary object: A new way to look at educational policy design and implementation. *Vocations and Learning, 2*(1), 19–35. https://doi.org/10.1007/s12186-008-9015-0

ENB. (2000a, April 17). *Earth negotiations bulletin -COP-11 Highlights: CITES-11)*, Vol. 21 No. 08. 1–2. Accessed Nov 25, 2012, from http://www.iisd.ca/vol21/enb2108e.html

ENB. (2000b, April 17). *Earth negotiations bulletin–COP-11 Highlights: 21(CITES-11), Vol. 21 No. 08. 1–2*. Accessed May 17, 2012, from http://www.iisd.ca/vol21/enb2108e.html

Estevez, E., & Janowski, T. (2013). Electronic governance for sustainable development—Conceptual framework and state of research. *Government Information Quarterly, 30*(Supplement 1(0)), S94–S109. https://doi.org/10.1016/j.giq.2012.11.001

Estévez-Abe, E. (2003). State -society partnership in the Japanese welfare state. In S. J. Pharr & F. J. Schwartz (Eds.), *The state of civil society in Japan* (pp. 154–170). Cambridge University Press.

Faloudi, A., & Waterhout, B. (2002). *The making of the European spatial development perspective: No masterplan, RTPI Library Series*. Routledge.

Feldman, D. L. (2012). The future of environmental networks—Governance and civil society in a global context. *Futures, 44*(9), 787–796. https://doi.org/10.1016/j.futures.2012.07.007

Force, U. B. (2014). *Emerging technologies: Smarter ways to fight wildlife crime*.

Fourteenth United Nations Congress on Crime Prevention and Criminal Justice, Kyoto, Japan 7-12 March 2021–WEMS. (n.d.). https://wems-initiative.org/event/fourteenth-united-nations-congress-on-crime-prevention-and-criminal-justice-kyoto-japan-7-12-march-2021/

Freedom House. (2012). *Freedom in the world 2012: The Arab uprisings and their global repercussions*. Accessed May 17, 2012, from http://www.freedomhouse.org/report/freedom-world/freedom-world-2012

Ganguly, S. (2010). *A constructivist analysis linking norm diffusion to policy networks*. Paper presented at the Conference on the human dimensions of global environmental change. Accessed Sep 5, 2012, from http://edocs.fuberlin.de/docs/servlets/MCRFileNodeServlet/FUDOCS_derivate_000000001331/Ganguly-A_constructivist_analysis_linking_norm_diffusion-269.pdf

Garland, E. (2008). The elephant in the room: Confronting the colonial character of wildlife conservation in Africa. *African Studies Review, 51*(03), 51–74. https://doi.org/10.1353/arw.0.0095

Gastil, J., & Levine, P. (2011). *Deliberative democracy handbook: Strategies for effective civic engagement in the 21st century*. John Wiley & Sons Inc.

Georgiadou, Y. (2008). *Using Geo-information from a market to a polis perspective*. Paper presented at the Inaugural Address, Enschede. Accessed Feb 5, 2011, from https://www.itc.nl/library/papers_2008/scie/georgiadou.pdf

Gibson, C. C. (1999). *Politicians and poachers: The political economy of wildlife policy in Africa*. Cambridge University Press.

Gieryn, T. F. (1983). Boundary-work and the demarcation of science from non-science: Strains and interests in professional ideologies of scientists. *American Sociological Review, 48*(6), 781–795. https://doi.org/10.2307/2095325

Gockeritz, S., Schultz, P. W., Rendon, T., Cialdini, R. B., Goldstein, N. J., & Griskevicius, V. (2010). Descriptive normative beliefs and conservation behaviour: The moderating roles of personal involvement and injunctive normative beliefs. *European Journal of Social Psychology, 40*(3), 514–523. https://doi.org/10.1002/ejsp.643

Gomar, J. O. V., & Stringer, L. C. (2011). Moving towards sustainability? An analysis of CITES' conservation policies. *Environmental Policy and Governance, 21*(4), 240–258. https://doi.org/10.1002/eet.577

Guha, R. (1999). *Environmentalism: A global history* (Vol. xiii). Longman.

Gupta, A. (2010). Transparency as contested political terrain: Who knows what about the global GMO trade and why does it matter? *Global Environmental Politics, 10*(3), 32–52. https://doi.org/10.1162/GLEP_a_00013

Hanson, R. (2004). *Japan-US impasse over lifting mad cow ban*. Asia Times.

Hardin, G. (1968). The tragedy of the commons. *Science, 162*(3859), 1243–1248. https://doi.org/10.1126/science.162.3859.1243

Henry, A. D., Lubell, M., & McCoy, M. (2010). Belief systems and social capital as drivers of policy network structure: The case of California regional planning. *Journal of Public Administration Research and Theory., 21*, 419. https://doi.org/10.1093/jopart/muq042

Hilborn, R., Arcese, P., Borner, M., Hando, J., Hopcraft, G., Loibooki, M., et al. (2006). Effective enforcement in a conservation area. *Science, 314*(5803), 1266. https://doi.org/10.1126/science.1132780

Hisschemöller, M., Hoppe, R., & Dunn, W. (2001). *Knowledge, power and participation in environmental policy analysis*. Transaction Publishers.

Hisschemöller, M., & Hoppe, R. (1995). Coping with intractable controversies: The case for problem structuring in policy design and analysis. *Knowledge, Technology & Policy, 8*(4), 40–60. https://doi.org/10.1007/bf02832229

Ho, P. (2001). Greening without conflict? Environmentalism, NGOs and civil society in China. *Development and Change, 32*(5), 893–921.

Hoffman, A. J., & Ventresca, M. J. (1999). The institutional framing of policy debates: Economics versus the environment. *American Behavioral Scientist, 42*(8), 1368–1392. https://doi.org/10.1177/00027649921954903

Hood, C. (2000). *The art of the state: Culture, rhetoric, and public management*. Oxford University Press.

Hoppe, R. (2007). Applied cultural theory as a tool in policy analysis. In F. Fischer, G. Miller, & M. S. Sidney (Eds.), *Handbook of public policy analysis: Theory, politics, and methods* (pp. 289–308). CRC/Taylor & Francis.

Hoppe, R. (2009). Scientific advice and public policy: Expert advisers' and policymakers' discourses on boundary work. *Poiesis & Praxis: International Journal of Technology Assessment and Ethics of Science, 6*, 235–263. https://doi.org/10.1007/s10202-008-0053-3

Hoppe, R. (2011). *The governance of problems: Puzzling, powering, participation*. Policy Press.

Hoppe, R., Wesselink, A., & Cairns, R. (2013). Lost in the problem: The role of boundary organisations in the governance of climate change. *Wiley Interdisciplinary Reviews: WIREs Climate Change, 2013*(4), 283–300. https://doi.org/10.1002/wcc.225

Hutton, J., & Dickson, B. (2000). CITES? Does it offer wild species a future? *Oryx, 34*(01), 1–2. https://doi.org/10.1046/j.1365-3008.2000.3ed341.x

Ingold, K., & Varone, F. (2011). Treating policy brokers seriously: Evidence from the climate policy. *Journal of Public Administration Research and Theory, 22*(2), 319–346. https://doi.org/10.1093/jopart/mur035

INTERPOL. (2014). *Environmental crime*. Accessed Jun 4, 2011, from http://www.interpol.int/Crime-areas/Environmental-crime/Environmental-crime

IUCN. (2000). *A report by IUCN–the world conservation union on the effectiveness of trade measures contained in the convention on international trade in endangered species of wild Fauna and Flora (CITES).* Trade measures in multilateral environmental agreements. Accessed May 7, 2012, from https://cites.org/common/prog/economics/iucn-trademeasuresinCITES.pdf

IUCN. (2014). *Summary statistics–trends in the status of biodiversity.* Accessed Jun 25, 2014, from http://www.iucnredlist.org/about/links

Jasanoff, S., & Martello, M. L. (2004). *Earthly politics: Local and global in environmental.* MIT Press.

Jasanoff, S., & Wynne, B. (1998). *Science and decision making human choice and climate change.* Batelle Press.

Jinnah, S. (2014). *Post-treaty politics: Secretariat influence in global environmental governance.* MIT Press.

Jones, M. (2012). Has CITES had its day? *BBC-Science & Environment.* Accessed Mar 7, 2012, from http://news.bbc.co.uk/2/hi/science/nature/8606011.stm

Jones, M. D., & McBeth, M. K. (2010). A narrative policy framework: Clear enough to be wrong? *Policy Studies Journal, 38*(2), 329–353.

Keck, M. E., & Sikkink, K. (1999). Transnational advocacy networks in international and regional politics. *International Social Science Journal, 51*(1), 89–101. https://doi.org/10.1111/1468-2451.00179

Lusaka Agreement Task Force. (2023). *Kenya assumes presidency as parties to Lusaka agreement convene extraordinary 13th Governing Council Meeting* Accessed Apr 14, 2025, from https://lusakaagreement.org/kenya-assumes-presidency-as-parties-to-lusaka-agreement-convene-extraordinary-13th-governing-council-meeting

Kenya, G. O. (2011). *Speech by his excellency hon. Mwai Kibaki, C.G.H., M.P., president–Republic of Kenya, during the African elephant Law enforcement day and ivory disposal ceremony held at Kenya wildlife service, Manyani field training school on 20th July, 2011. Nairobi: Government of Kenya.* Accessed Aug 12, 2012, from http://www.statehousekenya.go.ke/speeches/kibaki/july2011/2011200701.htm

Kretser, H. E., Wong, R., Roberton, S., Pershyn, C., Huang, J., Sun, F., & Zahler, P. (2014). Mobile decision-tree tool technology as a means to detect wildlife crimes and build enforcement networks. *Biological Conservation, 0*, 33. https://doi.org/10.1016/j.biocon.2014.08.018

Langdon, W. (1980). Do artifacts have politics? *Daedalus, 109*(1), 121–136. https://doi.org/10.2307/20024652

LATF. (2014). *Progress report on WEMS Africa implementation.* Lusaka Agreement Task Force.

LATF. (2015). *Summary report on seizure data in Wildlife Enforcement Monitoring System (WEMS).* Retrieved 12/7, 2015, from:http://lusakaagreement.org/wp-content/uploads/2015/04/Summary-WEMS-ANALYSIS-Report_2015.pdf [Accessed on 12/8/2011]

Levina, N., & Vaast, E. (2005). The emergence of boundary spanning competence in practice: Implications for implementation and use of information systems. *MIS Quarterly, 29*(2), 335–363. https://doi.org/10.2307/25148682

Levine, A. (2002). Convergence or convenience? International conservation NGOs and development assistance in Tanzania. *World Development, 30*(6), 1043–1055. https://doi.org/10.1016/s0305-750x(02)00022-0

Li, T. (2007). *The will to improve governmentality, development, and the practice of politics.* Accessed May 8, 2012, from http://public.eblib.com/EBLPublic/PublicView.do?ptiID=1170540

Lusaka Agreement Task Force. (2019). *Transboundary Information Sharing and Governance in Africa: The need for the hour.* Accessed Apr 14, 2025, from https://lusakaagreement.org/wp-content/uploads/2019/08/LATFSideEventatTICAD7-1.pdf

Marks, P. (2014). Elephants and rhinos benefit from drone surveillance. *New Scientist, 221*(2956), 24.

Mattson, D. J., Byrd, K. L., Rutherford, M. B., Brown, S. R., & Clark, T. W. (2006). Finding common ground in large carnivore conservation: Mapping contending perspectives. *Environmental Science & Policy, 9*(4), 392–405. https://doi.org/10.1016/j.envsci.2006.01.005

Mazmanian, D. A., & Sabatier, P. A. (1983). *Implementation and public policy*. Foresman.

Mazzucato, M., Eaves, D., & Vasconcellos, B. (2024). *Digital public infrastructure and public value: What is 'public' about DPI ? UCL Institute for Innovation and Public Purpose*, Working Paper Series (IIPP WP 2024–05). Available at: https://www.ucl.ac.uk/bartlett/public-purpose/publications/2024/mar/digital-public-infrastructure-and-public-value-what-public-about-dpi

Mcknight, B. (2007). Local understandings: Boundary objects in high conflict settings. In *Paper presented at the international Conference on organizational learning*.

Miles, M. B., & Huberman, A. M. (1999). *Qualitative data analysis: An expanded sourcebook*. Sage.

Mintrom, M., & Vergari, S. (1996). Advocacy coalitions, policy entrepreneurs, and policy change. *Policy Studies Journal, 24*(3), 420–434. https://doi.org/10.1111/j.1541-0072.1996.tb01638.x

Mitchell, R. B. (2003). International environmental agreements: A survey of their features, formation, and effects. *Annual Review of Environment and Resources, 28*(1), 429–461. https://doi.org/10.1146/annurev.energy.28.050302.105603

MOEF. (2005). *Joining the dots. Report of the tiger task force*. Accessed Sep 7, 2012, from http://www.projecttiger.nic.in/TTF2005/pdf/full_report.pdf

Mol, A. P. J. (2006). Environmental governance in the information age: The emergence of informational governance. *Environment and Planning C: Government and Policy, 24*(4), 17.

Moore, L. E. (2010). Conservation heroes versus environmental villains: Perceiving elephants in Caprivi, Namibia. *Human Ecology, 38*(1), 19–29. https://doi.org/10.1007/s10745-009-9290-x

Mrosovsky, N. (1997). IUCN'S credibility critically endangered. *Nature, 389*(6650), 436.

Muramatsu, M., & Krauss, E. S. (1984). Bureaucrats and politicians in policymaking: The case of Japan. *The American Political Science Review, 78*(1), 126–146.

Neff, M. W. (2011). What research should be done and why? Four competing visions among ecologists. *Frontiers in Ecology and the Environment, 9*(8), 462–469. https://doi.org/10.1890/100035

Neumann, R. P. (1995). Local challenges to global agendas–conservation, economic liberalization and the pastoralists rights movement in Tanzania. *Antipode, 27*(4), 363. https://doi.org/10.1111/j.1467-8330.1995.tb00285.x

Newmark, W., & Hough, J. (2000). Conserving wildlife in Africa: Integrated conservation and development projects and beyond, *BioScience, 50*(7), 585–592.

NGM. (2010). *South East Asia's Legal Wildlife Trade, 2014*. Accessed Nov 25, 2012, from http://ngm.nationalgeographic.com/2010/01/asian-wildlife/wildlife-graphic

Nijman, V. (2010). An overview of international wildlife trade from Southeast Asia. *Biodiversity and Conservation, 19*(4), 1101–1114. https://doi.org/10.1007/s10531-009-9758-4

Nijman, V., & Shepherd, C. R. (2011). The role of Thailand in the international trade in CITES-listed live reptiles and amphibians. *PLoS One, 6*(3), e17825. https://doi.org/10.1371/journal.pone.0017825

Niraj, S. K., Krausman, P. R., & Dayal, V. (2012). A stakeholder perspective into wildlife policy in India. *Journal of Wildlife Management, 76*(1), 10–18. https://doi.org/10.1002/jwmg.263

Nutley, S. M., Walter, I., & Davies, H. T. O. (2007). *Using evidence: How research can inform public*. Policy Press.

Orlikowski, W. J., & Barley, S. R. (2001). Technology and institutions: What can research on information technology and research on organizations learn from each other? *MIS Quarterly, 25*(2), 145–165.

Ostrom, E. (1991). Rational choice theory and institutional analysis: Toward complementarity. *The American Political Science Review, 85*(1), 237–243. https://doi.org/10.2307/1962889

Ostrom, E. (2009a). *Beyond markets and states: Polycentric governance of complex economic systems*. Prize Lecture workshop in 'Political Theory and Policy Analysis', Indiana University and Center for the Study of institutional diversity, Arizona State University. Accessed Apr 7, 2012, from http://www.nobelprize.org/nobel_prizes/economic-sciences/laureates/2009/ostrom_lecture.pdf

Ostrom, E. (2009b). *Understanding institutional diversity*. Princeton University Press.

Pan African Conference on Strengthening Information Sharing Infrastructure and Governance Frameworks to Address Human-Nature Conflicts. (2019). *WEMS initiative*. Accessed Apr 14, 2025, from https://wems-initiative.org/event/pan-african-conference-on-strengthening-information-sharing-infrastructure-and-governance-frameworks-to-address-human-nature-conflicts

Partridge, E. (1984). Nature as a moral resource. *Environmental Ethics, 6*, 101–130.

Phelps, J., Webb, E. L., Bickford, D., Nijman, V., & Sodhi, N. S. (2010). *Boosting CITES*. 330(6012), 1752–1753. https://doi.org/10.1126/science.1195558.

Pope. (2024). *Platformland: An anatomy of next-generation public services*. London School of Economics and Political Science.

Rastogi, A., Hickey, G. M., Badola, R., & Hussain, S. A. (2013). Diverging viewpoints on tiger conservation: A Q-method study and survey of conservation professionals in India. *Biological Conservation, 161*(0), 182–192. https://doi.org/10.1016/j.biocon.2013.03.013

Reeve, R. (2001). Verification mechanisms in CITES. *Verification Yearbook*, 137–156.

Reeve, R. (2002). Information systems. In *Policing international trade in endangered species: The CITES treaty and compliance* (p. 47). Earth Scan Publications.

Reeve, R. (2014). *Policing international trade in endangered species: The CITES treaty and compliance*. Taylor & Francis.

Richter, C. (2014, April 30). *Digital transformations in Indian cities: Between paper list and GIS map*, Enschede (PhD Thesis).

Robinson, C. J., & Wallington, T. J. (2012). Boundary work: Engaging knowledge Systems in co-management of feral animals on indigenous lands. *Ecology and Society, 17*(2). https://doi.org/10.5751/es-04836-170216

Rodrigues, A. S. L., Pilgrim, J. D., Lamoreux, J. F., Hoffmann, M., & Brooks, T. M. (2006). The value of the IUCN red list for conservation. *Trends in Ecology & Evolution, 21*(2), 71–76. https://doi.org/10.1016/j.tree.2005.10.010

Roozen, T. (1998). A case of need: The struggle to protect big-leaf mahogany. *Natural Resources Journal, 38*(4), 603–633.

Rosenzweig, P., & Van Weering, E. (2004). *WWF and the ivory trade*. IMD case study, 3, 19. Accessed Oct 25, 2011, from https://ecms.imd.ch/ecms/groups/cs_public/@i/@cs/@caid/@31269/@cimd_p133042/documents/cs_imd/cimd_p133042.hcst

Sabatier, P. A. (1988). An advocacy coalition framework of policy change and the role of policy-oriented learning therein. *Policy Sciences, 21*(2), 129–168. https://doi.org/10.1007/bf00136406

Sabatier, P. A. (Ed.). (2007). *Theories of the policy process* (2nd ed.). Westview Press.

Sabatier, P. A., & Jenkins-Smith, H. C. (1993). *Policy change and learning: An advocacy coalition approach*. Westview Press.

Salazar, D. J., & Alper, D. K. (2002). Reconciling environmentalism and the left: Perspectives on democracy and social justice in British Columbia's environmental movement. *Canadian Journal of Political Science (Revue Canadienne De Science Politique), 35*(3), 527–566.

Sandbrook, C. (2015). The social implications of using drones for biodiversity conservation. *Ambio, 44*(4), 636–647. https://doi.org/10.1007/s13280-015-0714-0

Sands, P. (2003). *Principles of international environmental Law*. Cambridge University Press.

Sato, J. (2003). Public land for the people: The institutional basis of community forestry in Thailand. *Journal of Southeast Asian Studies, 34*(02), 329–346. https://doi.org/10.1017/S0022463403000286

Schmolck, P. (2014). *The Q Method Page*. Accessed Jun 14, 2014, from http://schmolck.userweb.mwn.de/qmethod

Schut, M., van Paassen, A., & Leeuwis, C. (2013). Beyond the research–policy interface. Boundary arrangements at research–stakeholder interfaces in the policy debate on biofuel sustainability in Mozambique. *Environmental Science & Policy, 27*(0), 91–102. https://doi.org/10.1016/j.envsci.2012.10.007

Schwartz, F. (2002). Civil Society in Japan Reconsidered. *Japanese Journal of Political Science, 3*(02), 195–215.

Schwartz, F. J., & Pharr, S. J. (2003). *The state of civil society in Japan*. Cambridge University Press.

Sellar, J. (2013). *Just how bad is it?* Accessed Sep 5, 2013, from http://voices.news24.com/john-m-sellar/2013/08/just-how-bad-is-it/

Sewell, G. C. (2005). *Actors, coalitions and the framework convention on climate change*. (PhD Thesis). Institute of Technology.

Shirky, C. (2011). The political power of social media. *Foreign Affairs, 90*(1), 28–41.

Sickler, J., Fraser, J., Webler, T., Reiss, D., Boyle, P., Lyn, H., & Gruber, S. (2006). Social narratives surrounding dolphins: Q method study. *Society & Animals, 14*(4), 351–382.

Siebenhüner, B. (2009). The biodiversity secretariat: Lean shark in troubled waters. In *Managers of global change: The influence of international environmental bureaucracies*. MIT Press.

Smith, A. (2000). Policy networks and advocacy coalitions: Explaining policy change and stability in UK industrial pollution policy? *Environment and Planning C: Government and Policy, 18*, 95–114.

Snellen, I., Thaens, M., & Donk, W. V. (2012). *Public Administration in the Information age: Revisited* (p. 146). IOS Press BV.

Sodhi, N. S., et al. (2004). Southeast Asian biodiversity: An impending disaster. *Trends in Ecology & Evolution, 19*(12), 654–660.

Sodhi, N. S., et al. (2009). The state and conservation of southeast Asian biodiversity. *Biodiversity and Conservation, 19*(2), 317–328.

Spatial data infrastructure for Addressing Data Gaps in Combating Illegal Wildlife Trade and Meeting Sustainable Development Goals in Africa "A Call to Action" – WEMS. (2022). Accessed Apr 14, 2025, from https://wems-initiative.org/event/latf-cites-cop19/

Star, S. L., & Griesemer, J. R. (1989). Institutional ecology, 'translations and boundary objects: Amateurs and professionals in Berkeley's Museum of Vertebrate Zoology', 1907-39. *Social Studies of Science, 19*(3), 387–420. https://doi.org/10.2307/285080

Star, S. L., & Ruhleder, K. (1996). Steps toward an ecology of infrastructure: Design and access for large information spaces. *Information Systems Research, 7*(1), 111–134. https://doi.org/10.1287/isre.7.1.111

Steinberg, P. F. (2003). Understanding policy change in developing countries: The spheres of influence framework. *Global Environmental Politics, 3*(1), 11–32. https://doi.org/10.1162/152638003763336365

Stephenson, W. (1954). The study of behavior: q-technique and its methodology. [Book Review]. *Psychometrika, 19*(4), 327–333.

Stoett, P. (2002). The international regulation of trade in wildlife: Institutional and normative considerations. *International Environmental Agreements: Politics, Law and Economics, 2*(2), 193–208. https://doi.org/10.1023/a:1020942110468

Stone, D. A. (2002). *Policy paradox: The art of political decision making*. Norton.

Strauss, A. L., & Fagerhaugh, S. Y. (1997). *Social Organization of Medical Work*. Transaction Publishers.

Sukumar, R. (2003). *The living elephants: Evolutionary ecology, behaviour, and conservation*. Oxford University Press.

Swedlow, B. (2012). Cultural coproduction of four states of knowledge. *Science Technology & Human Values, 37*(3), 151–179. https://doi.org/10.1177/0162243911405345

Swinbanks, D. (1987). Japan takes action to control trade of endangered wildlife.' [Editorial Material]. *Nature, 326*(6114), 633–633.

The Second United Nations World Geospatial Information Congress Hyderabad, India, United Nations Statistics Division. (2022). *UNSD—UN-GGIM*. Accessed Apr 14, 2025, from https://ggim.un.org/2unwgic/

Thompson, M., Ellis, R., & Wildavsky, A. (1990). *Cultural theory*. Westview Press.

TICAD7 Side Event. (2019). *Lusaka agreement task force*. Accessed Apr 14, 2025, from https://lusakaagreement.org/ticad7-side-event

Tōgō, K. (2010). *Major Policy issues from 2004–2009*. in Japan's Foreign Policy, 1945–2009: The Quest for a proactive policy, pp. 403–432, Brill.

TRAFFIC. (2012). *EU-TWIX- Trade in Wildlife Information Exchange*. Accessed Sep 9, 2013, from http://www.eutwix.org/En/Default.aspx

UK, G. (2014). *Declaration: London Conference on the illegal wildlife trade*. Government of United Kingdom. Accessed Aug 9, 2014, from https://www.gov.UK/government/publications/declaration-london-conference-on-the-illegal-wildlife-trade

UN. (2025). *Universal safeguards for inclusive digital public infrastructure*. Accessed Apr 14, 2025, from https://www.dpi-safeguards.org/

UN. (1945). *Charter of the United Nations*. United Nations. Accessed Apr 5, 2013, from http://www.un.org/en/documents/charter/chapter10.shtml

UN. (2013). World Wildlife Day. A/C.2/68/L.48, Agenda item 19. Accessed Jan 1, 2014, from http://www.un.org/en/ga/search/view_doc.asp?symbol=A/C.2/68/L.48

UNDESA. (2012). *Results of survey of UN resident coordinators (RCs) and members of UN country teams (UNCTs)*. Preparation for the 2012 quadrennial comprehensive policy review of operational activities for development of the United Nations system 6 June, 2012. Accessed May 25, 2014, from http://www.un.org/esa/coordination/pdf/unct_survey_report_6_june_2012.pdf

UNEP. (2005). *A decade of regional wildlife Law enforcement: The case of the Lusaka agreement. [Report]*. United Nations Environmental Programme.

UNEP. (2007). 'Multilateral environmental agreement -negotiator's handbook 2007', Accessed Feb 14, 2012, from http://unfccc.int/resource/docs/publications/negotiators_handbook.pdf

UNEP. (2008). *UNEP yearbook 2008: An overview of our changing environment'*. United Nations environment Programme (UNEP). Accessed May 15, 2014, from http://www.unep.org/yearbook/2008/

UNEP-MONUSCO-OSESG. (2015). *Experts' background report on illegal exploitation and trade in natural resources benefitting organized criminal groups and recommendations on MONUSCO's role in fostering stability and peace in Eastern DR Congo Final report -April15th 2015*. Accessed Jun 20, 2015, from http://postconflict.unep.ch/publications/UNEP_DRCongo_MONUSCO_OSESG_final_report.pdf

United Nations Statistics Division. (2022). *UNSD—UN-GGIM*. https://ggim.un.org/2unwgic/documents/Remi%20Chandran.pdf

UNU. (2011). *Africa's war against wildlife crime continues*. Media Report. Accessed Jan 5, 2012, from http://unu.edu/news/news/africas-war-against-wildlife-crime-continues.html

UNU. (2014a). *Tokyo Conference on Combating International Wildlife Crime through Transboundary Information-Sharing* (Tokyo, 3–5 March, 2014) [Summary Report]. Accessed Dec 12, 2012, from http://wems-initiative.org/d1/unu%20summary%20of%20tokyo%20conference-combating%20int_wildl_crime.pdf

UNU. (2014b). *Invited participants*. Tokyo Conference on Combating Wildlife Crime. Accessed May 27, 2014, from http://wems-initiative.org/d1/tokyo%20conference%20bio%20speakers.pdf

Vandergeest, P. (1996). Property rights in protected areas: Obstacles to community involvement as a solution in Thailand. *Environmental Conservation, 23*(03), 259–268. https://doi.org/10.1017/S037689290003887X

Wasser, S. K., Clark, W. J., Drori, O., Kisamo, E. S., Mailand, C., Mutayoba, B., & Stephens, M. (2008). Combating the illegal trade in African elephant ivory with DNA forensics. *Conservation Biology, 22*(4), 1065–1071. https://doi.org/10.1111/j.1523-1739.2008.01012.x

Wasser, S. K., Mailand, C., Booth, R., Mutayoba, B., Kisamo, E., Clark, B., & Stephens, M. (2007). Using DNA to track the origin of the largest ivory seizure since the 1989 trade ban. *Proceedings of the National Academy of Sciences, 104*(10), 4228–4233. https://doi.org/10.1073/pnas.0609714104

Webler, T., Danielson, S., & Tuler, S. (2009). *Using Q method to reveal social perspectives in environmental research*. Accessed Apr 15, 2012, from http://www.seri-us.org/sites/default/files/Qprimer.pdf

Weible, C. M., & Sabatier, P. A. (2009). Coalitions, science, and belief change: Comparing adversarial and collaborative policy subsystems. *Policy Studies Journal, 37*(2), 195–212. https://doi.org/10.1111/j.1541-0072.2009.00310.x

Weible, C. M., Pattison, A., & Sabatier, P. A. (2010). Harnessing expert-based information for learning and the sustainable management of complex socio-ecological systems. *Environmental Science & Policy, 13*(6), 522–534. https://doi.org/10.1016/j.envsci.2010.05.005

Wellsmith, M. (2011). Wildlife crime: The problems of enforcement. *European Journal on Criminal Policy and Research, 17*(2), 125–148. https://doi.org/10.1007/s10610-011-9140-4

Wildavsky, A. (1987). Choosing preferences by constructing institutions: A cultural theory of preference formation. *The American Political Science Review, 81*(1), 4–21. https://doi.org/10.2307/1960776

Wolsink, M. (2004). Policy beliefs in spatial decisions: Contrasting core beliefs concerning space-making for waste infrastructure. *Urban Studies, 41*(13), 2669–2690. https://doi.org/10.1080/0042098042000294619

Worrall, A. (2012). *Running head: Boundary object theory*. Accessed July 7, 2012, from http://adamworrall4.files.wordpress.com/2012/07/6278_paper3_boundary_object_theory_analysis.pdf

WTO. (2001). *The Doha round*. Accessed May 26, 2013, from http://www.wto.org/english/tratop_e/dda_e/dda_e.htm#development

Young, S. (2003). Contemporary issues of the convention on international trade in endangered species of wild Fauna and Flora (CITES) and the debate over sustainable use. *Colorado Journal of International Environmental Law and Policy, 14*(1), 167–191.

Zhang, L., Hua, N., & Sun, S. (2008). Wildlife trade, consumption and conservation awareness in Southwest China. *Biodiversity and Conservation, 17*(6), 1493–1516. https://doi.org/10.1007/s10531-008-9358-8

Zurcher, S. (2005). Public participation in community forest policy in Thailand: The influence of academics as brokers. *Geografisk Tidsskrift-Danish Journal of Geography, 105*(1), 77–88.

The manufacturer's authorised representative in the EU is Springer Nature Customer Service Centre GmbH, Europaplatz 3, 69115 Heidelberg, Germany. If you have any concerns regarding our products, please contact ProductSafety@springernature.com

Printed and bound by CPI Group (UK) Ltd, Croydon, CR0 4YY
01/10/2026
02225954-0014